Spiritual Warfare from a Military Perspective

Boot Camp

A Bible Study Workbook

by Cristi Erickson

ISBNs
978-1-939944-56-6 Print
978-1-939944-57-3 Kindle
978-1-939944-58-0 ePub

Printed in the U.S.A.
First printing Dec. 2024
Published by Kingdom House Publishing, Lakebay, WA

Foreword

Terry Myers, pastor's wife of 60 years, Christian Women's speaker, and Bible teacher

Praise for SPIRITUAL WARFARE FROM A MILITARY PERSPECTIVE.

I firmly believe God has anointed my dear friend, Cristi Erickson, with a gift to communicate her heart to this generation of Christian believers, and spiritual warriors.

Cristi has had four years of active duty in the United States Marine Corps, she is a qualified educator which has included Bible teaching on many subjects. She is very well versed and appreciated by all under her tutorage!

In this study manual and workbook, she shares an interesting parallel of the warfare in a Christian's life and in military creeds, orders and commitment levels.

<u>The Soldier's Creed</u>

I will always place the mission first.
I will never accept defeat.
I will never quit.
I will never leave a fallen comrade.
I am a warrior and a member of a team.
I serve the people of the United States, and live the Army Values.
I am disciplined, physically and mentally tough, trained and proficient in my warrior tasks and drills.
I always maintain my arms, my equipment and myself.
I am an expert and I am a professional.
I stand ready to deploy, engage, and destroy the enemies of the United States of America in close combat.
I am a guardian of freedom and the American way of life.
I am an American Soldier.

Needless to say, if every Christian lived their Christianity by this Creed for God, there would be an unshakable foundation of Godliness!

This study of Spiritual Warfare from a military perspective, will challenge your heart to a fresh, new and deeper commitment to the Kingdom of God.

This commitment to lay down our lives for Christ and the Kingdom of God is life defining!

Romans 12:1
I beseech you therefore, brethren, by the mercies of God, that ye present your bodies a living sacrifice, holy, acceptable unto God, which is your reasonable service.

Acknowledgments

There are many people who have played a significant role in helping me bring this book to completion, and I want to express my heartfelt gratitude to them. First, I am deeply thankful to Debra Lurker, Vera Casale, and Nancy Finks for their faithful prayers over this book before I even began writing. Next, I am incredibly grateful to Bruce Cook, who guided me step by step through the publishing process. His meticulous attention to detail in reviewing my drafts, his thoughtful corrections, and his tireless efforts to alleviate as much stress as possible made this journey much smoother.

I also want to extend my appreciation to the countless others who encouraged and supported me in various ways—whether by providing feedback on early drafts, offering financial assistance, or simply cheering me on throughout this endeavor. There are too many names to list, but I am profoundly grateful to each and every one of you. A special thanks goes to Terry Myers, whose thoughtful words and support were instrumental and are featured in the Foreword of this book.

Finally, and most importantly, I give all thanks and praise to the Lord, who revealed to me how so many aspects of the military reflect profound spiritual truths. As Scripture says, "For now we see in a mirror dimly, but then face to face; now I know in part, but then I will know fully just as I also have been fully known." (1 Cor. 13:12)

Introduction

Welcome to the first book of a process. While in the early stages of writing this book, I realized the information was so vast it needed to be broken up into more than one book. This first book, in a workbook format, brings you alongside me in my own research as I link Scripture with many of the lessons taught in boot camp. This journey is deeply rooted in my lifelong fascination with warfare, a path less typical for a female but one that has profoundly shaped my perspective.

Since childhood, my interest in warfare has been evident. As a young girl, I gravitated towards war movies and GI Joes, shunning the usual girlish interests like Barbies or playing house. This fascination only grew with time. I have always enjoyed collecting knives, reading books that contained combat, and contemplating both fictional and historical war scenarios. Then, at age 21, I decided to take a significant step by joining the United States Marine Corps. Inspired by observing the military life within my family, I chose to start as an enlisted person before pursuing an officer's commission. This decision was heavily influenced by the path of my father, who earned the esteemed designation of Mustanger. He was revered and beloved by his peers and subordinates alike, a level of respect I attribute to his experience as a Mustanger.

When I joined the Corps, I had every intention of making it my career. However, I had already submitted my life to God, and it was not His plan for me to stay in the military. During my 3rd year of service He directed me to leave the military once my enlistment was up, to go back to college, and to study to be a missionary. Since God is my ultimate Commander in Chief, I followed His orders.

My fascination with warfare, my military-related experiences, and my relationship with Christ have all heavily influenced the creation of this book. As a matter of fact, one particular experience, which was the foundation of this book's idea, occurred during my last year in the Marine Corps. I was going to teach a Bible study on spiritual warfare based on insights God had revealed to me. A night or two before I was to teach, I was lying in bed waiting for sleep to take over when God began unveiling profound similarities between the military's 11 General Orders and our Christian lives. These revelations came so swiftly that I asked God to slow down so I could get a pencil and piece of paper to write them down. The next morning, I searched the Scriptures and found verses that supported these ideas.

In my studies, I've come to understand that we are spiritual beings in physical bodies. God originally intended us to live forever in perfect physical bodies unable to die, but Adam sinned and our bodies became imperfect. From that point, our view of ourselves has been more attuned to the temporary, physical condition. This makes it more difficult for us to recognize the spiritual world than the physical. Be that as it may, I believe that the physical realm often mirrors the spiritual. Because of this, we can use things that happen in the physical to help us understand what happens in the spiritual. Therefore, I am using physical experiences from the military to shed light on spiritual warfare.

I invite you now to join me in my research as I compare some of the lessons from boot camp with scriptural backing. Please note, this study won't always provide direct scriptural answers to every question posed. Instead, it encourages personal reflection, prayer, and independent scriptural study, aligning with the biblical call to be diligent in seeking truth (2 Tim. 2:15; Acts 17:11). While the New American Standard Bible (1995 translation) is primarily used here, I encourage you to refer to other translations, as cross-referencing can enhance understanding.

Table of Contents

Oath of Enlistment

All people who enter any US military branch must swear allegiance to the Constitution, which is the foundation of our country and her laws. The basic Oath of Enlistment is:

> *I,_______, do solemnly swear (or affirm) that I will support and defend the Constitution of the United States against all enemies, foreign and domestic; that I will bear true faith and allegiance to the same; and that I will obey the orders of the President of the United States and the orders of the officers appointed over me, according to the regulations and the Uniform code of Military Justice. So help me God.*[1]

For officers, the wording is a little different, but they also swear an oath to "support and defend the Constitution of the United States against all enemies, foreign and domestic" and that they "will bear true faith and allegiance to the same."

What does this mean in a spiritual connotation? When we are "saved" or "born again," what is it we do? First, we repent of our sins, then we give our lives to Jesus (whom we have accepted to be God in the flesh), believing He died for our sins and has risen from the dead. So, ❶ we swear allegiance to God, turning our backs on (repenting of) our sinful "civilian" ways, and giving our life to Him and His cause; ❷ we promise to fight against all enemies, those from outside ourselves (spiritual attacks and outside influences) as well as those from within (habits and temptations); and ❸ we make a commitment to obey God (our ultimate authority) and submit to others in authority (Eph. 5:21) according to the document our laws are built on – the Bible. Finally, just like with the oath of military enlistment, the phrase "So help me God" resonates in our oath to God. We can fulfill our oath only with God's help.

This Oath of Enlistment is taken prior to going to basic training. Our minds and hearts are set towards our new goal – becoming a Marine, a Soldier, a Sailor, a Coast Guardsman, an Airman, or a Guardian. In the spiritual, our goal is to become Christlike, to become like the One who rules over this "military" group. With this goal we enter the rigors of boot camp (Basic Training), except in the Christian's life, boot camp and the real world happen at the same time. Sometimes the real battles hit hard right away, and sometimes they sneak in slowly. Though there is this difference between the military and the Christian worlds, we believers have the Holy Spirit to help us survive the attacks, and even to grow in them, while we learn the basics.

It has been said, and is often repeated, that the job of the drill instructors in Basic Training is to break you in order to remake you. A nicer way of putting this is that their job is to change your habits and your way of thinking so that every fiber of your being, including your thought processes, has totally absorbed and fully projects the military image so well that it requires little to no thought to act and react in a military manner. For the Christian, the difficulties that come our way are our drill instructors. It is the difficulties both in boot camp and in the Christian realm that mold us into the people we are to become, physically, emotionally, mentally, and spiritually. Do all military drill instructors succeed in their job? There are some recruits who do not graduate. I would not call these failures, but neither are they successes, for the rigors imposed by the drill instructors affect each recruit in some way whether they get all the way through boot camp or not. On the other hand, every recruit who

---

[1]"The US Military Oath of Enlistment." *Military.com*, 2024,< www.military.com/join-armed-forces/swearing-in-for-military-service.html>, Feb. 2024

graduates from boot camp is a success in varying degrees. I say "in varying degrees" because there are those who will look the part, and act the part, but have not fully aligned their thinking. Some will still opt for laziness or rebellion in some form or another. Also, there are some who have already given themselves to a higher authority (God), and they might be fully engaged in and accepting of the military way, but only as long as it does not go against conscience. I do not speak judgment here. I merely point out that no matter how much training or resolve, in the end we are individuals who make our own choices.

With all this in mind, to develop the kind of soldier that is needed, it makes perfect sense to develop the thought patterns and to learn the basics before going on to other endeavors, called Military Occupational Specialties (such as Truck Mechanic, Supply Clerk, or Pilot), especially when we remember that the main purpose of any U.S. military branch is to protect and defend against all enemies. There are many things taught in basic training, and I've decided to focus on eight of them.

Chapter 1: Chain of Command

In the military, we learn (and memorize) our Chain of Command all the way up to the President of the United States. The President, the highest authority, is also known as the Commander in Chief, and to have an audience with him, one would have to get approval from a great many people, knowing that any one of them could veto the request. Christians have a great advantage in their Chain of Command from what the US military has. Our Commander in Chief is God, our Father.

1. How do we gain access to our Commander in Chief?

   - What does Jesus mean by "through Me" as quoted in John 14:6?
     Jesus said to him, "I am the way, and the truth, and the life; no one comes to the Father but through Me."

   - What does Romans 10:9 tell us we need to do, and what has been done for us?
     that if you confess with your mouth Jesus as Lord, and believe in your heart that God raised Him from the dead, you will be saved;

   - What does Ephesians 2:8-9, 13 tell us we need to do, and what has been done for us?
     For by grace you have been saved through faith; and that not of yourselves, it is the gift of God; not as a result of works, so that no one may boast.... But now in Christ Jesus you who formerly were far off have been brought near by the blood of Christ.

2. What is our relationship with our Commander in Chief?

   - Ephesians 2:19
     So then you are no longer strangers and aliens, but you are fellow citizens with the saints, and are of God's household,

3. Why can we be confident that we will be able to speak to the Commander in Chief?

   - According to Ephesians 3:11-12, who gives us our confidence?
     This was in accordance with the eternal purpose which He carried out in Christ Jesus our Lord, in whom we have boldness and confident access through faith in Him.

- Hebrews 4:14-16 shows us why we can be confident. What is our reason?
  Therefore, since we have a great high priest who has passed through the heavens, Jesus the Son of God, let us hold fast our confession. For we do not have a high priest who cannot sympathize with our weaknesses, but One who has been tempted in all things as we are, yet without sin. Therefore, let us draw near with confidence to the throne of grace, so that we may receive mercy and find grace to help in time of need.

4. How does our Commander in Chief communicate with us?
   - Hebrews 1:1-2 (There is more than one way God communicates. Can you think of some examples from the Old Testament? If so, write the verses and summarize them.)
     God, after He spoke long ago to the fathers in the prophets in many portions and in many ways, in these last days has spoken to us in His Son, whom He appointed heir of all things, through whom also He made the world.

     - How do the methods of communication from "long ago" relate to the methods "in these last days"? Why is it better to be "in these last days?"

5. In the military, we can contact the Commander in Chief via a letter (which, by itself, is not likely to gain us access) or through the chain of command (our Commanding Officer, to his CO, to his CO, and so on until our request arrives at the top). Once we finally have access to the President, we address him with utmost respect and never in a familiar manner. How do we communicate with our Heavenly Commander in Chief and how would we address Him?
   - According to 1 Timothy 2:5-6, what did our mediator do to enable us to communicate?
     For there is one God, and one mediator also between God and men, the man Christ Jesus, who gave Himself as a ransom for all, the testimony given at the proper time.
   - According to Romans 8:26, who else helps us with our communication?

In the same way the Spirit also helps our weaknesses; for we do not know how to pray as we should, but the Spirit Himself intercedes for us with groanings too deep for words.

- In John 16:23, what do you think it means to "ask in Jesus's name"?
  In that day you will not question Me about anything. Truly, truly, I say to you, if you ask the Father for anything in My name, He will give it to you.

- In Matthew 6:9, how are we told to address our Heavenly Commander in Chief?
  Our Father who is in heaven, hallowed be Your name.

6. Consider this scenario: The U.S. President is your father, and you are in the military. You need to talk with the President concerning some urgent issue. How would you gain access to him? Would you go up the long ladder of the chain of command which could take a very long time (or be denied) or would you call him on the phone and ask for a meeting?

- Once you are in the meeting with only him, would you address him as Mr. President, or would you address him as Father (or Dad, Pop, etc.)?

- This is similar to our situation with our Heavenly Father thanks to the work Jesus did to gain us access to the Almighty God and to put us in a family relationship with Him.
  - What condition were we in before, what is our status now, and how do we keep this status?
    And although you were formerly alienated and hostile in mind, engaged in evil deeds, yet He has now reconciled you in His fleshly body through death, in order to present you before Him holy and blameless and beyond reproach – if indeed you continue in the faith firmly established and steadfast, and not moved away from the hope of the gospel that you have heard, which was proclaimed in all creation under heaven, and of which I, Paul, was made a minister. (Colossians 1:21-23)

- What is our status according to Romans 8:15, and what does it give us the right to address God as?
  For you have not received a spirit of slavery leading to fear again, but you have received a spirit of adoption as sons by which we cry out, "Abba! [1] *Father!"*

---

[1] According to Chad Herrington of Him Publications, "Abba" implies both intimacy and an obedient heart. (https://himpublications.com/blog/meaning-abba/) This fits the idea of a soldier whose father happens to be President of the United States. That soldier will likely address his father in private as "Dad," but will also have respect for him and be willing to obey him as his Commander in Chief.

Chapter 2: The UCMJ and the Code of Conduct

The UCMJ (Uniform Code of Military Justice) is the base upon which is built the US military laws. While in boot camp, we are taught the basics, but we do not go through a lot of details. All service personnel, when they are active duty (even temporarily), are under these laws. As Christians, we are always active duty, and we are always under the laws of God. The UCMJ determines the legal course to take when something wrong is done. Likewise, the Bible tells us the ramifications of our sin and of continual intentional sin. It also describes how our Judge will respond based on certain parameters and how it is that we should behave. I will not get into the details of either the UCMJ or the many Scriptures that can be used for the Lord's judicial system since this Bible study is about warfare, not justice. However, these are part of the life of every soldier; therefore, it is our duty to read, study, and know them.

The U.S. Military Code of Conduct is a guide with six articles which lay out the desired behaviors and mindset for all military personnel. To follow this guide, each person must understand the principles and general ideas presented as well as the words that are stated. For example, the first paragraph of Article 1 in this Code states, "I am an American, fighting in the forces which guard my country and our way of life. I am prepared to give my life in their defense." (USMC Marine Corps Institute) This states who I am, what my main purpose is, and how far I'm willing to go to achieve my purpose.

For the Christian, the Bible is our Code of Conduct, giving us the instructions for our behavior and our perspective. Reading it can help us understand the principles and general ideas it presents. Studying and meditating on it can help us personalize these principles and lessons and make them a part of us in our thoughts and actions. Memorizing it can help put its words deep in our hearts and minds, enabling us to recall them when we face certain situations that call for the responses they encourage or command.

Code of Conduct, Article I, paragraph 1

> *I am an American, fighting in the forces which guard my country and our way of life. I am prepared to give my life in their defense.*[1]

1. Who does this paragraph say "I" am?

2. What does it say my purpose is?

---

[1] USMC Marine Corps Institute, "The United States Marines – Essential Subjects" (Arlington: USMC, Dept. of the Navy, 1983) pages 1-1 through 1-3, and additional Code of Conduct Articles are on pages 1-3 through 1-14. *This source is the same for the remaining Code of Conduct Articles cited in this chapter, so it will not be repeated elsewhere.*

3. What does it call me to do?

4. In Ephesians 2:10, who does it say we are?
   For we are His workmanship, created in Christ Jesus for good works, which God prepared beforehand so that we would walk in them"

   - What does it say our purpose is?

   - What does it call us to do?

Code of Conduct, Article II, paragraph 1

> *I will never surrender of my own free will. If in command, I will never surrender the members of my command while they have the means to resist.*

5. According to Ephesians 6:12, who or what are we fighting against? Write your answer below the verse.
   For our struggle is not against flesh and blood, but against the rulers, against the powers, against the world forces of this darkness, against the spiritual forces of wickedness in the heavenly places.

   - Is this battle physical or spiritual? How do you know?

   - Why is it important to keep in mind whether this is a physical or spiritual battle?

If you are having trouble with the "Why" question, consider that physical battles and spiritual battles have different armies engaged in them. A spiritual battle may encroach upon the physical, but the true warriors are spiritual; the physical are only "pawns."

6. In looking at this Code of Conduct article, who are the two peoples which I am not to surrender?

    -
    -

7. Matthew 23:4 gives an example of how "commanding officers" might surrender their units to the enemy.
    They tie up heavy burdens and lay them on men's shoulders, but they themselves are unwilling to move them with so much as a finger.
    - What are these leaders doing that is wrong?
    - What should they do instead? If you can, support your idea with Scripture.

8. Though this Code of Conduct article mentions the situation where I might be in command, who does Galatians 6:1 say should restore those who had surrendered, but want freedom? How should this be done?
    Brethren, even if anyone is caught in any trespass, you who are spiritual, restore such a one in a spirit of gentleness; each one looking to yourself, so that you too will not be tempted.

9. Describe spiritual people according to the following verses.
    - *For the mind set on the flesh is death, but the mind set on the Spirit is life and peace.* (Romans 8:6)
    - *But a natural man does not accept the things of the Spirit of God, for they are foolishness to him; and he cannot understand them, because they are spiritually appraised. But he who is spiritual appraises all things, yet he himself is appraised by no one.* (1 Corinthians 2:14-15)

- (These verses describe those who are fleshly, so answer with the opposite.)
  And I, brethren, could not speak to you as to spiritual men, but as to men of flesh, as to infants in Christ. I gave you milk to drink, not solid food; for you were not yet able to receive it. Indeed, even now you are not yet able, for you are still fleshly. For since there is jealousy and strife among you, are you not fleshly, and are you not walking like mere men? (1 Corinthians 3:1-3)

10. We are all called to do what we can to help our fellow soldiers to resist the enemy (to not surrender). What are some ways we can do this? Bear in mind that our own actions often give encouragement and strength to others to do likewise.

    - *Keep watching and praying that you may not enter into temptation; the spirit is willing, but the flesh is weak.* (Matthew 26:41 & Mark 14:38)

    - *Let no one seek his own good, but that of his neighbor.* (1 Corinthians 10:24)

    - *For you were called to freedom, brethren; only do not turn your freedom into an opportunity for the flesh but through love serve one another.* (Galatians 5:13)

    - *This command I entrust to you, Timothy, my son, in accordance with the prophecies previously made concerning you, that by them you fight the good fight, keeping faith and a good conscience, which some have rejected and suffered shipwreck in regard to their faith.* (1 Timothy 1:18-19)

    - *Fight the good fight of faith; take hold of the eternal life to which you were called, and you made the good confession in the presence of many witnesses. I charge you in the presence of God, who gives life to all things, and of Christ Jesus, who testified the good confession before Pontius Pilate, that you keep the commandment without stain or reproach until the appearing of our Lord Jesus Christ.* (1 Timothy 6:12-14)

    - *For God is not unjust so as to forget your work and the love which you have shown toward His name, in having ministered and in still ministering to the saints. And we desire that each one of you show the same diligence so as to realize the full assurance of hope until the end, so that you will not be sluggish, but imitators of those who through faith and patience inherit the promises.* (Hebrews 6:10-12)

- *Therefore, strengthen the hands that are weak and the knees that are feeble, and make straight paths for your feet, so that the limb which is lame may not be put out of joint, but rather be healed.* (Hebrews 12:12-13)

- *Therefore, humble yourselves under the mighty hand of God, that He may exalt you at the proper time, casting all your anxiety on Him, because He cares for you. Be of sober spirit, be on the alert. Your adversary the devil, prowls around like a roaring lion, seeking someone to devour. But resist him, firm in your faith, knowing that the same experiences of suffering are being accomplished by your brethren who are in the world.* (1 Peter 5:6-9)

I know that people have a habit in Bible studies to just try to answer the question by using words from the particular verses given. I ask you now to go back and look again at question 16 and at the verses I've given, and consider specific situations and examples that could occur in your life. I also ask that you discipline yourself from this point on to truly consider the questions, verses, and answers in a way that makes them personal to you.

11. *"...while they have the means to resist."* Do we ever **not** have the means to resist? Sometimes a person or a squad can become so physically injured that they cannot fight back. Can this happen spiritually? (You'll find the answer <u>after</u> fully answering question 18, so do question 18, then come back to 17.)

12. How is it we can resist?

- *One of your men puts to flight a thousand, for the LORD your God is He who fights for you, just as He promised you.* (Joshua 23:10)

- *And if one can overpower him who is alone, two can resist him. A cord of three strands is not quickly torn apart.* (Ecclesiastes 4:12)

- *By smooth words he will turn to godlessness those who act wickedly toward the covenant, but the people who know their God will display strength and take action.* (Daniel 11:32)

- *No temptation has overtaken you but such as is common to man; and God is faithful, who will not allow you to be tempted beyond what you are able, but with the temptation will provide the way of escape also, so that you will be able to endure it.* (1 Corinthians 10:13)

- *Seeing that His divine power has granted to us everything pertaining to life and godliness, through the true knowledge of Him who called us by His own glory and excellence. For by these He has granted to us His precious and magnificent promises, so that by them you may become partakers of the divine nature, having escaped the corruption that is in the world by lust.* (2 Peter 1:3-4)

- *Submit therefore, to God. Resist the devil and he will flee from you.* (James 4:7)

Now, go back and answer question 17, then proceed to Article III, paragraph 1.

Code of Conduct, Article III, paragraph 1

> *If I am captured I will continue to resist by all means available. I will make every effort to escape and to aid others to escape. I will accept neither parole nor special favors from the enemy.*

13. According to Romans 3:23, who has been captured?
 For all have sinned and fall short of the glory of God.

14. In John 8:34, what does Jesus say has captured us or continually tries to capture us?

Jesus answered them, "Truly, truly, I say to you, everyone who commits sin is the slave of sin."

15. 2 Peter 2:18-19 sheds light on some who try to bait us into slavery, and he uses the pronoun "they." Read the following verses, then answer the questions that follow.

    For speaking out arrogant words of vanity they entice by fleshly desires, by sensuality, those who barely escape from the ones who live in error, promising them freedom while they themselves are slaves of corruption; for by what a man is overcome, by this he is enslaved. (2 Pet. 2:18-19)

    Then the Lord knows how to rescue the godly from temptation and to keep the unrighteous under punishment for the day of judgment, and especially those who indulge the flesh in its corrupt desires and despise authority. (2 Pet. 2:9-10)

    - Who are "they?"

    - How do they trick us back into slavery (2 Pet. 2:18-19)?

    - Are Christians exempt from this bondage? Explain your answer.
      For if, after they have escaped the defilements of the world by the knowledge of the Lord and Savior Jesus Christ, they are again entangled in them and are overcome, the last state has become worse for them than the first. For it would be better for them not to have known the way of righteousness, than having known it, to turn away from the holy commandment handed on to them. (2 Peter. 2:20-21)

        - Are you wondering if you can trust Christian leaders? How can you be sure they are not leading you into slavery? The following verses tell how. After reading each, write who is responsible for this verification and how it is to be done.

            1. *Be diligent to present yourselves approved to God as a workman who does not need to be ashamed, accurately handling the word of truth.* (2 Timothy 2:15)

2. *Now these were more noble-minded than those in Thessalonica, for they received the word with great eagerness, examining the Scriptures daily to see whether these things were so.* (Acts 17:11)

16. Why should we resist even if we have been captured?

- *What shall we say then? Are we to continue in sin so that grace may increase? May it never be! How shall we who died to sin still live in it?* (Romans 6:1-2)

- *But resist him, firm in your faith, knowing that the same experiences of suffering are being accomplished by your brethren who are in the world. After you have suffered for a little while, the God of all grace, who called you to His eternal glory in Christ, will Himself perfect, confirm, strengthen and establish you.* (1 Peter 5:9-10)

17. The captured military person needs a particular mindset in order to continue to resist and to seek escape from their imprisonment. This takes both physical and mental discipline to focus on the goal of freedom and not to beat oneself up for getting captured. In fact, the soldier must continually remind himself that he is a free man who just happens to be in temporary bondage, and he must look and plan for any opportunity to free himself. He must keep in his mind the feelings and pictures of what freedom is or he will succumb to a slave's mentality and become comfortable in his current state. To survive (and escape) one needs hope. What do these verses tell us about the way we should be thinking?

- *Therefore, there is now no condemnation for those who are in Christ Jesus"* (Romans 8:1)

- *For those who are according to the flesh set their minds on things of the flesh, but those who are according to the Spirit, the things of the Spirit. For the mind set on the flesh is death, but the mind set on the Spirit is life and peace, because the mind set on the flesh is hostile toward God; for it does not subject itself to the law of God, for it is not even able to do so, and those who are in the flesh cannot please God.* (Romans 8:5-12)

- *For I consider that the sufferings of this present time are not worthy to be compared with the glory that is to be revealed to us. For the anxious longing of the creation waits eagerly for the revealing of the sons of God. For the creation was subjected to futility, not willingly, but because of Him who subjected it, in hope that the creation itself also will be set free from its slavery to corruption into the freedom of the glory of the children of God. For we know that the whole creation groans and suffers the pains of childbirth together until now. And not only this, but also we ourselves, having the first fruits of the Spirit, even we ourselves groan within ourselves, waiting eagerly for our adoption as sons, the redemption of our body. For in hope we have been saved, but hope that is seen is not hope; for who hopes for what he already sees? But if we hope for what we do not see, with perseverance we wait eagerly for it.* (Romans 8:18-25)

- *Therefore, be careful how you walk, not as unwise men but as wise, making the most of your time, because the days are evil.* (Ephesians 5:15-16)

Accepting parole would be accepting a short term or permanent freedom from the prison. The prisoner could be offered the ability to go on a trip somewhere or to go home. Special favors could be things like extra food or blankets, medical care for himself or a buddy, or being put on a special work detail that is less demanding. All of these offers, of course, would be for a "small" cost such as signing a paper which has lies on it or becoming an informant against his fellow prisoners.

18. Why would someone accept parole or special favors? How do you think this would affect other prisoners of war?

19. For a Christian, accepting "parole or special favors" is like trying to live in 2 camps at the same time. It might be done because it seems an easier path (less persecution and/or more familiar). What do these verses tell you about this?

- *He who walks with wise men will be wise, but the companion of fools will suffer harm.* (Proverbs 13:20)

- *No one can serve two masters; for either he will hate the one and love the other, or he will be devoted to one and despise the other. You cannot serve God and wealth.* (Matthew 6:24)

- *And do not be conformed to this world, but be transformed by the renewing of your mind, so that you may prove what the will of God is, that which is good and acceptable and perfect.* (Romans 12:2)

- *For in the case of those who have once been enlightened and have tasted of the heavenly gift and have been made partakers of the Holy Spirit, and have tasted the good word of God and the powers of the age to come, and then have fallen away, it is impossible to renew them again to repentance, since they again crucify to themselves the Son of God and put Him to open shame. For ground that drinks the rain which often falls on it and brings forth vegetation useful to those for whose sake it is also tilled, receives a blessing from God; but if it yields thorns and thistles, it is worthless and close to being cursed, and it ends up being burned.* (Hebrews 6:4-8)

- *Do not love the world nor the things in the world. If anyone loves the world, the love of the Father is not in him. For all that is in the world, the lust of the flesh and the lust of the eyes and the boastful pride of life, is not from the Father, but is from the world. The world is passing away, and also its lusts; but the one who does the will of God lives forever.* (1 John 2:15-17)

- *Let no one say when he is tempted, 'I am being tempted by God'; for God cannot be tempted by evil, and He Himself does not tempt anyone. But each one is tempted when he is carried away and enticed by his own lust. Then when lust has conceived, it gives birth to sin; and when sin is accomplished, it brings forth death.* (James 1:13-15)

20. If we "accept parole or special favors," how would our actions affect us, our fellow believers, and/or others? Consider this question and write any insight as you read the following verses.

- *"Woe to the rebellious children" declares the LORD, "who execute a plan, but not Mine, and make an alliance, but not of My Spirit, in order to add sin to sin; who proceed down to Egypt without consulting Me, to take refuge in the safety of Pharaoh and to seek shelter in the shadow of Egypt! Therefore, the safety of Pharaoh will be your shame and the shelter in the shadow of Egypt, your humiliation."* (Isaiah 30:1-3)

- *Now we who are strong ought to bear the weaknesses of those without strength and not just please ourselves. Each of us is to please his neighbor for his good, to his edification.* (Romans 15:1-2)

- *But it was because of the false brethren secretly brought in, who had sneaked in to spy out our liberty which we have in Christ Jesus, in order to bring us into bondage. But we did not yield in subjection to them for even an hour, so that the truth of the gospel would remain with you.* (Galatians 2:4-5)

- *But when Cephas came to Antioch, I opposed him to his face, because he stood condemned. For prior to the coming of certain men from James, he used to eat with the Gentiles; but when they came, he began to withdraw and hold himself aloof, fearing the party of the circumcision. The rest of the Jews joined him in hypocrisy, with the result that even Barnabas was carried away by their hypocrisy.* (Galatians 2:11-13)

- *And keep a good conscience so that in the thing in which you are slandered, those who revile your good behavior in Christ will be put to shame.* (1 Peter 3:16)

- *Do not be deceived: "Bad company corrupts good morals."* (1 Corinthians 15:33)

- *But whoever causes one of these little ones who believe in Me to stumble, it would be better for him to have a heavy millstone hung around his neck, and to be drowned in the depth of the sea.* (Matthew 18:6)

As soldiers of Christ, we must take care to not give in to temptation, and especially to not continue giving in to that temptation. This habitual sin becomes a trap called being "caught up in a sin." If we have been caught up in a sin, we need to do what we can to free ourselves from that sin. At the same time, we should not seek the way of least resistance, for that is not the path God has called us to. For our own sake, we must not compromise our walk with Christ, for it is very difficult to free oneself from the entanglements that come with sin, especially when it is voluntarily done in spite of knowing it is wrong. Additionally, for the sake of our brothers and sisters in Christ, we need to remain as pure as we can. We strengthen ourselves and each other through what we experience – our experiences when we have resisted, and those in which we have fallen, then overcome.

21. Code of Conduct, Article IV, paragraph 1

> *If I become a prisoner of war, I will keep faith with my fellow prisoners. I will give no information or take part in any action which might be harmful to my comrades. If I am senior, I will take command. If not, I will obey the lawful orders of those appointed over me and will back them up in every way.*

To keep faith with someone means to continue supporting or being loyal to them. I have already touched on this with the section on Article III when referencing the idea of not accepting parole or special favors. Not taking part in any action which might be harmful is included in this concept. Therefore, let's look at the rest of this article.

The person who is "senior" is the person who should have authority. In the military, it would be the person with the highest rank. After that person, it would be the next highest-ranking person, and so on all the way down to the lowest rank (in which case it would be whoever had been at that rank the longest).

For civilian Christians in America, it might make no sense to think that we would be affected by the statement, "If I am senior, I will take command." After all, we can attend a church, and that would most likely make us NOT the leader in our church community; the pastor or reverend would be senior. Leadership within the Church Body actually depends upon individuals – their calling, their skills, their walk with God, and in some cases, the voting results of the congregation or specific members of the congregation. However, what about within the family? Does Scripture explain how leadership works within the family?

22. According to these verses, what is the order of seniority within a family unit and what are the responsibilities of each level of authority? Read the following verses before you answer.

- *But I want you to understand that Christ is the head of every man, and the man is the head of a woman, and God is the head of Christ.* (1 Corinthians 11:3)

- *Wives, be subject to your own husbands, as to the Lord. For the husband is the head of the wife, as Christ also is the head of the church, He Himself being the Savior of the body. But as the church is subject to Christ, so also the wives ought to be to their husbands in everything.* (Ephesians 5:22-24)

- *Fathers, do not provoke your children to anger, but bring them up in the discipline and instruction of the Lord.* (Ephesians 6:4)

- *Wives, be subject to your husbands, as is fitting in the Lord.* (Colossians 3:18)

- *Children, obey your parents in the Lord, for this is right.* (Ephesians 6:1)

- *Children, be obedient to your parents in all things, for this is well-pleasing to the Lord.* (Colossians 3:20)

- *In the same way, you wives, be submissive to your own husbands so that even if any of them are disobedient to the word, they may be won without a word by the behavior of their wives.* (1 Peter 3:1)

23. The Code of Conduct states that subordinates are to obey lawful orders. Therefore, we must understand what is lawful and right so that we can understand when an order is not lawful. This shows that the one with seniority also has responsibilities, not just the right to give commands. How does that work in the Biblical family? What responsibility does the leader in the family have and what are the parameters of that responsibility?

    - *Husbands, love your wives, just as Christ also loved the church and gave Himself up for her,* (Ephesians 5:25)

    To fully understand the husband's responsibility as it relates to Eph. 5:25, you would need to understand how Christ showed His love. Not only did He sacrifice His life for us, He sacrificed His time to listen, to teach, and to encourage. He put others' needs ahead of His own, and He put the will of God the Father before His own will.

    - *So, husbands ought also to love their own wives as their own bodies. He who loves his own wife loves himself;* (Ephesians 5:28)

There are several examples in the Old Testament of men who were not good fathers because they didn't discipline their children well. Here are a couple of examples. What did these fathers do that was wrong?

- *Now Eli was very old; and he heard all that his sons were doing to all Israel, and how they lay with the women who served at the doorway of the tent of meeting...Then a man of God came to Eli and said to him, "Thus says the LORD, 'Did I not indeed reveal Myself to the house of your father when they were in Egypt in Bondage to Pharaoh's house? Did I not choose them from all the tribes of Israel to be My priests, to go up to My altar, to burn incense, to carry an ephod before Me; and did I not give to the house of your father all the fire offerings of the sons of Israel? Why do you kick at My sacrifice and at My offering which I have commanded in My dwelling, and honor your sons above Me, by making yourselves fat with the choicest of every offering of My people Israel?'"* (1 Samuel 2:22, 27-29)

- *Now Adonijah the son of Haggith exalted himself, saying, "I will be king." So he prepared for himself chariots and horsemen with fifty men to run before him. His father had never crossed him at any time by asking, "Why have you done so?"* (1 Kings 1:5-6) [David was the king at this time, but he was very old and was unaware of this latest thing that Adonijah had done. Prior to this, David had promised Bathsheba that their son, Solomon, would be his successor.]

24. How are we to view and behave towards others within the faith community (also called our church family)?

    - *Do not sharply rebuke an older man, but rather appeal to him as a father, to the younger men as brothers, the older women as sisters, in all purity.* (1 Timothy 5:1-2) (Also, when you answer this, remember your answers from #28.)

25. Write this challenge to the head of the family in your own words.

    But if anyone does not provide for his own, and especially for those of his household, he has denied the faith and is worse than an unbeliever. (1 Timothy 5:8)

26. If you wonder about how our government fits in, look at these verses and answer who we are to obey.

    - *Every person is to be in subjection to the governing authorities. For there is no authority except from God, and those which exist are established by God.* (Romans 13:1)

    - *Submit yourselves for the Lord's sake to every human institution, whether to a king as the one in authority, or to governors as sent by him for the punishment of evildoers and the praise of those who do right. For such is the will of God that by doing right you may silence the ignorance of foolish men.* (1 Peter 2:13-15)

    - What is the exception?
      But Peter and the apostles answered, "We must obey God rather than men." (Acts 5:29)

        - How do you think the exception would relate to our line of authority?

27. What does it mean within family units and within the church to "take command," as Article IV states is the job of the one who is senior? Does it mean "My way or the highway" or is there more to it? Answer this question after reading the following verses:

    - *Train up a child in the way he should go, even when he is old he will not depart from it.* (Proverbs 22:6)

    - *and be subject to one another in the fear of Christ.... Husbands, love your wives, just as Christ also loved the church and gave Himself up for her.* (Ephesians 5:21, 25)

    - *Husbands, love your wives and do not be embittered against them.... Fathers, do not exasperate your children, so that they will not lose heart.* (Colossians 3:19,21)

    - *Remember those who led you, who spoke the word of God to you; and considering the result of their conduct, imitate their faith.* (Hebrews 13:7)

- *You husbands in the same way, live with your wives in an understanding way, as with someone weaker, since she is a woman; and show her honor as a fellow heir of the grace of life, so that your prayers will not be hindered.* (1 Peter 3:7)

- *You younger men, likewise, be subject to your elders; and all of you, clothe yourselves with humility toward one another, for God is opposed to the proud, but gives grace to the humble.* (1 Peter 5:5)

- *My little children, I am writing these things to you so that you may not sin. And if anyone sins, we have an Advocate with the Father, Jesus Christ the righteous.* (1 John 2:1)

According to Article IV, the lower-ranking member is to "*obey the lawful orders of those appointed over me and will back them up in every way.*" Note that this states that the higher-ranking commands should be "lawful orders." For the soldier, this means the commands must align with the Constitution, the Code of Conduct, and the Uniform Code of Military Justice. For the Christian, this means that they should align with Scripture. The action required is "obey," and what is obeyed is the scripturally-sound commands given by the recognized authority.

28. What is the responsibility of the lower-ranking member of the family or Christian group?

- *Therefore, be imitators of God, as beloved children.* (Ephesians 5:24)

- *Honor your father and mother (which is the first commandment with a promise), so that it may be well with you, and that you may live long on the earth.* (Ephesians 6:2-3)

If you live in a household in which there is no father/husband, remember the rule that the military has. WHOEVER is of the highest rank is the one in command. Therefore, in such a situation, the mother/wife will need to take upon herself the duties and responsibilities the father/husband would have had until such time as someone of "higher rank" becomes a part of that family. Additionally, remember that it is God who has the highest rank in any family, for He is the ultimate Father (John 1:12).

Code of Conduct, Article V, paragraph 1

> *When questioned, should I become a prisoner of war, I am required to give name, rank, service number, and date of birth. I will evade answering*

further questions to the utmost of my ability. I will make no oral or written statements disloyal to my country and its allies or harmful to their cause.

The reason for the few things I am required to give is so that my country can know it truly is me who has been captured. More information could give the enemy knowledge that could be used against other soldiers (captured or still fighting), especially when coupled with other information the enemy may have obtained. There is a valuable saying which came out of WWII that could be used to address this concept. It is "Loose lips sink ship," and it basically means to beware of saying more than you should. For the Christian, this can easily equate to gossip and other harmful speech.

29. What do these verses say about people who sin with their lips or about such sins?

- *You destroy those who speak falsehood; the LORD abhors the man of bloodshed and deceit. There is nothing reliable in what they say; their inward part is destruction itself. Their throat is an open grave; they flatter with their tongue.* (Psalm 5:6, 9)

- *They speak falsehood to one another; with flattering lips and with a double heart they speak.* (Psalm 12:2)

- *But to the wicked God says, "What right have you to tell of My statutes and to take My covenant in your mouth? You let your mouth loose in evil and your tongue frames deceit. You sit and speak against your brother; you slander your own mother's son."* (Psalm 50:16, 19-20)

- *He who goes about as a talebearer reveals secrets, but he who is trustworthy conceals a matter.* (Proverbs 11:13)

- *A perverse man spreads strife, and a slanderer separates intimate friends.* (Proverbs 16:28)

- *He who goes about as a slanderer reveals secrets, Therefore, do not associate with a gossip.* (Proverbs 20:19)

- *He who rebukes a man will afterward find more favor than he who flatters with the tongue.* (Proverbs 28:23)

- *"Behold, I am against the prophets," declares the LORD, "who use their tongues and declare, 'The Lord declares.' Behold, I am against those who have prophesied false dreams," declares the LORD, "and related them and led My people astray by their falsehoods and reckless boastings; yet I did not send them or command them, nor do they furnish this people the slightest benefit," declares the LORD.* (Jeremiah 23:31-32)

- *And just as they did not see fit to acknowledge God any longer, God gave them over to a depraved mind, to do those things which are not proper, being filled with all unrighteousness, wickedness, greed, evil; full of envy, murder, strife, deceit, malice; they are gossips, slanderers, haters of God, insolent, arrogant, boastful, inventors of evil, disobedient to parents, without understanding, untrustworthy, unloving, unmerciful; and although they know the ordinance of God, that those who practice such things are worthy of death, they not only do the same, but also give hearty approval to those who practice them.* (Romans 1:28-32)

30. What do these verses say about people's speech?

- *The Spirit of the LORD spoke by me, and His word was on my tongue.* (2 Samuel 23:2)

- *Even before there is a word on my tongue, behold, O LORD, You know it all.* (Psalm 139:4)

- *He who conceals hatred has lying lips, and he who spreads slander is a fool. When there are many words, transgression is unavoidable, but he who restrains his lips is wise. The tongue of the righteous is as choice silver, the heart of the wicked is worth little. The lips of the righteous feed many, but fools die for lack of understanding.* (Proverbs 10:18-21)

- *Death and life are in the power of the tongue, and those who love it will eat its fruit.* (Proverbs 18:21)

- *The Lord GOD has given me the tongue of disciples, that I may know how to sustain the weary one with a word. He awakens me morning by morning, He awakens my ear to listen as a disciple.* (Isaiah 50:4)

31. What do these verses say about how we should speak?

- *For as long as life is in me, and the breath of God is in my nostrils, my lips certainly will not speak unjustly, nor will my tongue mutter deceit.* (Job 27:3-4)

- *O LORD, who may abide in Your tent? Who may dwell on Your holy hill? He who walks with integrity, and works righteousness, and speaks truth in his heart. He does not slander with his tongue, nor does evil to his neighbor, nor takes up a reproach against his friend.* (Psalm 15:1-3)

- *Keep your tongue from evil and your lips from speaking deceit.* (Psalm 34:13)

- *I said, "I will guard my ways that I may not sin with my tongue; I will guard my mouth as with a muzzle while the wicked are in my presence."* (Psalm 39:1)

- *Let my cry come before You, O LORD; Give me understanding according to Your word. Let my supplication come before You; deliver me according to Your word. Let my lips utter praise, for You teach me Your statutes. Let my tongue sing of Your word, for all Your commandments are righteousness.* (Psalm 119:169-172)

- *He who restrains his words has knowledge, and he who has a cool spirit is a man of understanding. Even a fool, when he keeps silent, is considered wise; when he closes his lips, he is considered prudent.* (Proverbs 17:27-28)

- *He who guards his mouth and his tongue guards his soul from troubles.* (Proverbs 21:23)

- *An excellent wife, who can find? For her worth is far above jewels. She opens her mouth in wisdom, and the teaching of kindness is on her tongue.* (Proverbs 31:10, 26)

- *If anyone thinks himself to be religious, and yet does not bridle his tongue but deceives his own heart, this man's religion is worthless.* (James 1:26)

- *But no one can tame the tongue; it is a restless evil and full of deadly poison. With it we bless our Lord and Father, and with it we curse men, who have been made in the likeness of God; from the same mouth come both blessing and cursing. My brethren, these things ought not to be this way.* (James 3:8-10)

- *For, "The one who desires life, to love and see good days, must keep his tongue from evil and his lips from speaking deceit."* (1 Peter 3:10)

It might seem that verses like James 1:26 and James 3:8 contradict each other, but they do not. Just remember promises like in Matthew 19:26, *"With people this is impossible, but with God all things are possible,"* and Paul's comment in Philippians 4:13, *"I can do all things through Him who gives me strength."*

The last sentence in the first paragraph of Article V states, *"I will make no oral or written statements disloyal to my country and its allies or harmful to their cause."* What does this mean to the Christian? Let's dissect this together.

32. In the US Code of Conduct, our country is America. The way to get home to America after being captured is either to successfully escape or for our country to free us somehow (a prisoner exchange, win the war, etc.). What do the following verses tell us about our country (our kingdom) as Christians and how we get there?

- *Not everyone who says to Me, "Lord, Lord," will enter the kingdom of heaven, but he who does the will of My Father who is in heaven will enter.* (Matthew 7:21)

- *Jesus answered and said to him, "Truly, truly, I say to you, unless one is born again he cannot see the kingdom of God."* (John 3:3)

- *For He rescued us from the domain of darkness, and transferred us to the kingdom of His beloved Son, in whom we have redemption, the forgiveness of sins.* (Colossians 1:13)

- *The Lord will rescue me from every evil deed, and will bring me safely to His heavenly kingdom; to Him be the glory forever and ever, Amen.* (2 Timothy 4:18)

Notice in 2 Tim. 4:18 and Col. 1:13, it is God who rescues us. We cannot escape on our own power. It is somewhat like a prisoner swap because Jesus had our sins put on Himself. What was swapped was the sin. At the same time, it isn't like a prisoner swap, but more like a rescue because Jesus never belonged to the enemy and He didn't stay a prisoner either. On His own merit, He actually was never captured since He never sinned.

- *Therefore, since we receive a kingdom which cannot be shaken, let us show gratitude, by which we may offer to God an acceptable service with reverence and awe.* (Hebrews 12:28)

- *Listen, my beloved brethren: did not God choose the poor of this world to be rich in faith and heirs of the kingdom which He promised to those who love Him?* (James 2:5)

33. We have citizens in our spiritual country, called Christian believers, and we have allies of our country. Who might these allies be according to Romans 11:1-24?

    I say then, God has not rejected His people, has He? May it never be! For I too am an Israelite, a descendant of Abraham, of the tribe of Benjamin. God has not rejected His people whom He foreknew. Or do you not know what the Scripture says in the passage about Elijah, how he pleads with God against Israel? "Lord, they have killed your prophets, they have torn down your altars, and I alone am left, and they are seeking my life. But what is the divine response to him? "I have kept for Myself seven thousand men who have not bowed the knee to Baal." In the same way then, there has also come to be at the present time a remnant according to God's gracious choice. But if it is by grace, it is no longer on the basis of works, otherwise grace is no longer grace.

    What then? What Israel is seeking, it has not obtained, but those who were chosen obtained it, and the rest were hardened; just as it is written, "God gave them a spirit of stupor, eyes to see not and ears to hear not, down to this very day." And David says, "Let their table become a snare and a trap, and a stumbling block and a retribution to them. Let their eyes be darkened to see not, and bend their backs forever."

    I say then, they did not stumble so as to fall, did they? May it never be! But by their transgression salvation has come to the Gentiles, to make them jealous. Now if their transgression is riches for the world and their failure is riches for the Gentiles, how much more will their fulfillment be! But I am speaking to you who are Gentiles. Inasmuch then as I am an apostle of Gentiles, I magnify my ministry, if somehow I might move to jealousy my fellow countrymen and save some of them. For if their rejection is the reconciliation of the world, what will their acceptance be but life from the dead? If the first piece of dough is holy, the lump is also; and if the root is holy, the branches are too.

    But if some of the branches were broken off, and you, being a wild olive, were grafted in among them and became partaker with them of the rich root of the olive tree, do not be arrogant toward the branches; but if you are arrogant, remember that it is not you who supports the root, but the root supports you.

    You will say then, 'Branches were broken off so that I might be grafted in.' Quite right, they were broken off for their unbelief, but you stand by your faith. Do not be conceited, but fear; for if God did not spare the natural branches, He will not spare you, either. Behold then the kindness and severity of God; to those who fell, severity, but to you, God's kindness, if you continue in His kindness; otherwise you also will be cut off. And they also, if they do not continue in their unbelief, will be grafted in, for God is able to graft them in again. For if you were cut off from what is by nature a wild olive tree, and were grafted contrary to nature into a cultivated olive tree, how much more will these who are the natural branches be grafted into their own olive tree.

34. So, what would it look like to say or write something that would harm our country and her people or our allies? What is the most harmful thing we could do?

- Thinking of our "allies" as described in Romans 11, and what we're told in Article V (page 24), what should we <u>not</u> do according to Leviticus 19:14? *"You shall not curse a deaf man, nor place a stumbling block before the blind, but you shall revere your God; I am the LORD."*

    - How would John 9:39 relate to Leviticus 19:14?
    And Jesus said, "For judgment I came into this world, so that those who do not see may see, and that those who see may become blind." (John 9:39)

- We are not to be like the world. 2 Corinthians 4:4 tells us what the god of this world has done. How should we behave towards non-believers in light of what Jesus has done? (Consider Leviticus 19:14 again.)
"in whose case the god of this world has blinded the minds of the unbelieving so that they might not see the light of the gospel of the glory of Christ, who is the image of God."

- How should we behave towards our fellow citizens (fellow-believers)?

    - *Therefore, let us not judge one another anymore, but rather determine this – not to put an obstacle or a stumbling block in a brother's way.* (Romans 14:13)

    - *But take care that this liberty of yours does not somehow become a stumbling block to the weak. For if someone sees you, who have knowledge, dining in an idol's temple, will not his conscience, if he is weak, be strengthened to eat things sacrificed to idols? For through your knowledge he who is weak is ruined, the brother for whose sake Christ died. And so, by sinning against the brethren and wounding their conscience when it is weak, you sin against Christ. Therefore, if food causes my brother to stumble, I will never eat meat again, so that I will not cause my brother to stumble.* (1 Corinthians 8:9-13)

Code of Conduct, Article VI, paragraph 1

> *I will never forget that I am an American, fighting for freedom, responsible for my actions, and dedicated to the principles which made my country free. I will trust in my God and the United States of America.*

When I think of this paragraph's first phrase, "I will never forget that I am an American," I recall a poem I memorized long ago called "A Cross in My Pocket" by Verna Mae Thomas. This poem was either sold with a little cross or, in my case, a little cross was glued to a small piece of wallet-sized paper which had the poem on it.

I carry a cross in my pocket,
A simple reminder to me
Of the fact that I am a Christian
No matter where I may be.

This little cross is not magic,
Nor is it a good luck charm.
It isn't meant to protect me
From every physical harm.

It's not for identification
For all the world to see.
It's simply an understanding
Between my Savior and me.

When I put my hand in my pocket
To bring out a coin or a key,
The cross is there to remind me
Of the price He paid for me.

It reminds me, too, to be thankful
For my blessings day by day,
And to strive to serve Him better
In all that I do and say.

It's also a daily reminder
Of the peace and comfort I share
With all who know my Master
And give themselves to His care.

So, I carry a cross in my pocket,
Reminding no one but me

That Jesus Christ is Lord of my life,
If only I'll let Him be.

The "cross in my pocket" is a physical reminder of a spiritual truth. It reminds the carrier of who he is, how he got there, his main purpose, some ways to achieve that purpose, and that he is not alone. Article VI is a written reminder of who we (military persons) are, what our main purpose is, and how we are to achieve and uphold our role. Let's break it down with Scriptural support.

35. "*...I am an American...*" Who are you? Answer this question after each verse.

- *But now, thus says the LORD, your Creator, O Jacob, and He who formed you, O Israel, "Do not fear, for I have redeemed you; I have called you by name; you are Mine!"* (Isaiah 43:1)

- *But he is a Jew who is one inwardly; and circumcision is that which is of the heart, by the Spirit, not by the letter; and his praise is not from men, but from God.* (Romans 2:29)

- *But as many as received Him, to them He gave the right to become children of God, even to those who believe in His name, who were born, not of blood nor of the will of the flesh nor of the will of man, but of God.* (John 1:12-13)

- *So Jesus was saying to those Jews who had believed Him, "If you continue in My word, then you are truly disciples of Mine."* (John 8:31)

- *that if you confess with your mouth Jesus as Lord, and believe in your heart that God raised Him from the dead, you will be saved.* (Romans 10:9)

36. "*...fighting for freedom...*" What do these verses say about your purpose?

- *The LORD has made everything for its own purpose, even the wicked for the day of evil. The mind of a man plans his way, but the LORD directs his steps.* (Proverbs 16:4, 9)

- *And we know that God causes all things to work together for good to those who love God, to those who are called according to His purpose. For those whom He foreknew, He also predestined to become conformed to the image of His Son, so that He would be the firstborn among many brethren;* (Romans 8:28-29)

- *For indeed while we are in this tent, we groan, being burdened, because we do not want to be unclothed but to be clothed, so that what is mortal will be swallowed up by life. Now He who prepared us for this very purpose is God, who gave to us the Spirit as a pledge. Therefore, being always of good courage, and knowing that while we are at home in the body we are absent from the Lord – for we walk by faith, not by sight – we are of good courage, I say, and prefer rather to be absent from the body and to be at home with the Lord. Therefore, we also have as our ambition, whether at home or absent, to be pleasing to Him.* (2 Corinthians 5:4-9)

- *But have nothing to do with worldly fables fit only for old women. On the other hand, discipline yourselves for the purpose of godliness;* (1 Timothy 4:7)

- *For what credit is there if, when you sin and are harshly treated, you endure it with patience? But if when you do what is right and suffer for it you patiently endure it, this finds favor with God.* (1 Peter 2:20-21)

- *Therefore, since Christ has suffered in the flesh, arm yourselves also with the same purpose, because he who has suffered in the flesh has ceased from sin, so as to live the rest of the time in the flesh no longer for the lusts of men, but for the will of God.* (1 Peter 4:1-2)

37. We are in a fight whether we like it or not. What is our purpose or desired results of the fight?

    - *It was for freedom that Christ set us free; Therefore, keep standing firm and do not be subject again to a yoke of slavery. For you were called to freedom, brethren; only do not turn your freedom into an opportunity for the flesh, but through love serve one another.* (Galatians 5:1, 13-14)

- *This command I entrust to you, Timothy, my son, in accordance with the prophecies previously made concerning you, that by them you fight the good fight, keeping faith and a good conscience, which some have rejected and suffered shipwreck in regard to their faith.* (1 Timothy 1:18-19)

- *But flee from these things, you man of God, and pursue righteousness, godliness, faith, love, perseverance and gentleness. Fight the good fight of faith; take hold of the eternal life to which you were called, and you made the good confession in the presence of many witnesses.* (1 Timothy 6:11-12)

- *I have fought the good fight, I have finished the course, I have kept the faith;* (2 Timothy 4:7)

- *For speaking out arrogant words of vanity they entice by fleshly desires, by sensuality, those who barely escape from the ones who live in error, promising them freedom while they themselves are slaves of corruption; for by what a man is overcome, by this he is enslaved.* (2 Peter 2:18-19)

38. *"...responsible for my actions..."* How do you achieve your purpose?
Part 1: Where does responsibility lie and who is responsible to whom?

- *Be gracious to me, O God, according to Your lovingkindness; According to the greatness of Your compassion blot out my transgressions. Wash me thoroughly from my iniquity and cleanse me from my sin. For I know my transgressions, and my sin is ever before me. Against You, You only, I have sinned and done what is evil in Your sight, so that You are justified when You speak and blameless when You judge.* (Psalm 51:1-4)

- *A prudent man sees evil and hides himself, the naïve proceed and pay the penalty.* (Proverbs 27:12)

- *Whatever you do, do your work heartily, as for the Lord rather than for men, knowing that from the Lord you will receive the reward of the inheritance. It is the Lord Christ whom you serve.* (Colossians 3:23-24)

- *For we have become partakers of Christ, if we hold fast the beginning of our assurance firm until the end, while it is said, "Today if you hear His voice, do not harden your hearts, as when they provoked Me." For who provoked Him when they had heard? Indeed, did not all those who came out of Egypt led by Moses?* (Hebrews 3:14-16)

- *Therefore, to one who knows the right thing to do and does not do it, to him it is sin.* (James 4:17)

39. *"...dedicated to the principles which made my country free."* How do you achieve your purpose?
    Part 2: What made you free and where is your heart (dedication) to be? What actions show where your heart is?

    - *But know that the LORD has set apart the godly man for Himself; the LORD hears when I call to Him. Tremble, and do not sin; meditate in your heart upon your bed, and be still. Selah. Offer the sacrifices of righteousness, and trust in the LORD.* (Psalm 4:3-5)

    - *O LORD, who may abide in Your tent? Who may dwell on Your holy hill? He who walks with integrity, and works righteousness, and speaks truth in his heart. He does not slander with his tongue, nor does evil to his neighbor, nor takes up a reproach against his friend; in whose eyes a reprobate is despised, but who honors those who fear the LORD; he swears to his own hurt and does not change; he does not put out his money at interest, nor does he take a bribe against the innocent. He who does these things will never be shaken.* (Psalm 15:1-5)

    - *The law of the LORD is perfect, restoring the soul; the testimony of the LORD is sure, making wise the simple. The precepts of the LORD are right, rejoicing the heart; the commandment of the LORD is pure, enlightening the eyes. The fear of the LORD is clean, enduring forever; the judgments of the LORD are true; they are righteous altogether. They are more desirable than gold, yes, than much fine gold; sweeter also than honey and the drippings of the honeycomb. Moreover, by them Your servant is warned; in keeping them there is great reward.* (Psalm 19:7-11)

    - *My son, keep my words and treasure my commandments within you. Keep my commandments and live, and my teaching as the apple of your eye. Bind them on your fingers; write them on the tablet of your heart.* (Proverbs 7:1-3) What is implied by the locations where the words are to be placed (eyes, fingers, & heart)?

- *"What commandment is the foremost of all?" Jesus answered, "The foremost is, 'Hear o Israel! The Lord our God is one Lord; and you shall love the Lord your God with all your heart, and with all your soul, and with all your mind, and with all your strength.' The second is this, 'You shall love your neighbor as yourself.' There is no other commandment greater than these."* (Mark 12:29-31)

- *So if the Son makes you free, you will be free indeed.* (John 8:36)

- *that if you confess with your mouth Jesus as Lord, and believe in your heart that God raised Him from the dead, you will be saved; for with the heart a person believes, resulting in righteousness, and with the mouth he confesses, resulting in salvation.* (Romans 10:9-10)

- *Therefore, I urge you, brethren, by the mercies of God, to present your bodies a living and holy sacrifice, acceptable to God, which is your spiritual service of worship. And do not be conformed to this world, but be transformed by the renewing of your mind, so that you may prove what the will of God is, that which is good and acceptable and perfect.* (Romans 12:1-2)

- *Do nothing from selfishness or empty conceit, but with humility of mind regard one another as more important than yourselves; do not merely look out for your own personal interests, but also for the interests of others. Have this attitude in yourselves which was also in Christ Jesus* (Philippians 2:3-5)

40. *"I will trust in my God and in the United States of America."* Where should you place your trust, and if the verse has a contrast (what you should NOT trust in), what is it? (Psalm 44:6-8 has contrasts.)

    - *And those who know Your name will put their trust in You, for You, O LORD, who dwells in Zion; declare among the peoples His deeds.* (Psalm 9:10)

- *But I have trusted in Your lovingkindness; my heart shall rejoice in Your salvation.* (Psalm 13:5)

- *In You our fathers trusted; they trusted and You delivered them.* (Psalm 22:4-5)

- *Commit your way to the LORD, trust also in Him, and He will do it.* (Psalm 37:5)

- *For I will not trust in my bow, nor will my sword save me. But You have saved us from our adversaries, and You have put to shame those who hate us. In God we have boasted all day long, and we will give thanks to Your name forever. Selah.* (Psalm 44:6-8)

- *In God, whose word I praise, in God I have put my trust; I shall not be afraid. What can mere man do to me?* (Psalm 56:4)

- *Trust in the LORD with all your heart and do not lean on your own understanding. In all your ways acknowledge Him, and He will make your paths straight.* (Proverbs 3:5-6)

- *Nebuchadnezzar responded and said, "Blessed be the God of Shadrach, Meshach and Abed-nego, who has sent His angel and delivered His servants who put their trust in Him, violating the king's command, and yielded up their bodies so as not to serve or worship any god except their own God."* (Daniel 3:28)

In summary, the 6 Articles of the U.S. Military Code of Conduct are for all members of the Armed Forces of the United States. To live up to the Code, each person needs to know the ideologies and the concepts behind each Article as well as know the words. Knowing is not enough, though. Each person must take the responsibility upon himself to live out the principles and ideas incorporated within these Articles.

Like members of the Armed Forces, Christians often enter their new life with little to no knowledge of the words of the Code of Conduct (the Bible) nor of the belief system behind the words. These things are learned over time, but learning can be sped up by memorizing and studying the Scriptures. At the same time, practical application is the best way to make them your own personal principles.

Review of Code of Conduct as It Pertains to Christians

- **Article I** p. 9 *(Who am I & what is my purpose)*
  - I am God's workmanship
  - My purpose is to do good works through Christ
- **Article II** p. 10 *(Never surrender)*
  - Our true fight is in the spiritual realm and against spiritual forces.
  - I should never give in to or give up fighting against these spiritual forces or my own fleshly desires.
  - I should continue to help my fellow believers to overcome these enemies.
  - It is the Lord who gives us everything we need to avoid being captured and to win.
- **Article III** p. 14 *(If captured – try to escape, don't accept favors)*
  - When I've given in to sin, I shouldn't give up fighting to free myself from its entanglements and to live as an overcomer because, with God's help, I will win.
  - I should never allow myself to think that I've already gone so far that going a little further won't matter.
  - Once I am aware that something I do or want to do is a sin, I need to avoid it and fight not to give in to it. If I don't, it would be the same as a prisoner accepting parole or special favors.
- **Article IV** p. 20 *(If prisoner – keep faith w/comrades. Obey command structure.)*
  - I need to be aware that my actions will affect other Christians, even if I don't think they notice. Therefore, I must always seek to make godly choices, even when nobody is around.
  - Remember the chain of command:
    - The Father is over Jesus, Christ is over the Church, and the pastor is over the flock.
    - God is over each member of the family and the family as a whole.
    - The father is over the wife, and the parents (father and mother) are over the children.
  - I must maintain a humble mindset and be obedient to those in a higher position than I (including if they are not yet a believer), the exception being if they want me to do something that is against God's will for me.
- **Article V** p. 24 *(Prisoner – when questioned, no disloyalty)*
  - My tongue was made to praise God, not to degrade His creations. Gossip and lying are not something that God approves of. I must do my best to not partake in either – neither be a speaker of them nor a listener to those who do them.
  - Only God can help me control my tongue, so I need to lean heavily on Him in attempting to reign it in. With much practice and with never relenting on working on it, through Him I can learn to speak only upright and true words.
    - If my whole being is focused on God, my lips will learn to speak only as He wishes.

- **Article VI** p. 32 *(Never forget who I am. Trust in God & country)*
    - No matter where I am or who can see me, I must always remember that I am a representative of Jesus Christ. Therefore, I must always behave in a way that brings Him glory and portrays Him rightly.
    - It is beneficial to frequently remind myself of what Jesus has done for me and what He has freed me from. This will help me keep the right mindset to persevere through hardships and to remain humble enough to enable others to see Christ in and through me.

My heart is to be fully dedicated to Christ at every moment of every day. If my heart is His, my actions will align with His will. Also, if my eyes are focused on Him, I will not be distracted by the things around me.

Chapter 3: First Aid, NBC Defense, and Fitness

Fitness

Maintaining one's health or even improving it can ward off a lot of issues, so fitness is an important aspect of First Aid and NBC Defense. Fitness involves the body, mind, and emotions. This statement is not just for the military person, but also for the Christian. We must not neglect our physical health.

Physical, emotional, and mental fitness are closely bound to each other. Some agree that emotional and mental health will affect physical health, yet don't see as clearly how the reverse is also true. Even less often will people recognize that spiritual fitness and physical fitness affect each other. Let's take a closer look at the effects of physical health and at what affects it.

> "The responsibility of maintaining a satisfactory level of physical fitness is shared by each Marine and his commanding officer. The time required for the maintenance of these standards is traditional in the Marine Corps way-of-life and the necessary self-discipline that must be part of the character of all Marines. A program of regular, vigorous and progressive physical fitness training results in an increase in work efficiency, self-confidence, and personal as well as unit pride. A Marine who is not physically fit is a burden to the readiness and combat efficiency of his unit and detracts from the professionalism of the Marine Corps." (page 71 of *The United States Marine Essential Subjects*)

1. From the following verses, what do you learn about physical fitness/health and how it relates to things we think and do or allow?

    - *When I kept silent about my sin, my body wasted away through my groaning all day long.* (Psalm 32:3)

    - *Do not be wise in your own* eyes; *fear the LORD and turn away from evil. It will be healing to your body and refreshment to your bones.* (Proverbs 3:7-8)

    - *A tranquil heart is life to the body, but passion is rottenness to the bones.* (Proverbs 14:30)

- *But beyond this, my son, be warned: the writing of many books is endless, and excessive devotion to books is wearying to the body.* (Ecclesiastes 12:12)

- *The eye is the lamp of your body; when your eye is clear, your whole body also is full of light; but when it is bad, your body also is full of darkness.* (Luke 11:34)

2. From the following verses, what do you learn about the relationship to physical fitness/health and the way we think (our mental and spiritual fitness), and what do they tell you about how we should treat the physical body?

    - *Therefore, do not let sin reign in your mortal body so that you obey its lusts, and do not go on presenting the members of your body to sin as instruments of unrighteousness; but present yourselves to God as those alive from the dead, and your members as instruments of righteousness to God.* (Romans 6:12)

    - *But I see a different law in the members of my body, waging war against the law of my mind and making me a prisoner of the law of sin which is in my members.* (Romans 7:23)

    - *For if you are living according to the flesh, you must die; but if by the Spirit you are putting to death the deeds of the body, you will live.* (Romans 8:13)

    - *Or do you not know that your body is a temple of the Holy Spirit who is in you, whom you have from God, and that you are not your own? For you have been bought with a price: Therefore, glorify God in your body.* (1 Corinthians 6:19-20)

    - *Everyone who competes in the games exercises self-control in all things. They then do it to receive a perishable wreath, but we an imperishable. Therefore, I run in such a way, as not without aim; I box in such a way, as not beating the air; but I discipline my body and make it my slave, so that, after I have preached to others, I myself will not be disqualified.* (1 Corinthians 9:25-27)

- *Therefore, consider the members of your earthly body as dead to immorality, impurity, passion*[1]*, evil desire, and greed, which amounts to idolatry.* (Colossians 3:5)

3. What danger do you see in the following verse for the Christian who seeks to do what is right?

   - *If you have died with Christ to the elementary principles of the world, why, as if you were living in the world, do you submit yourself to decrees, such as, "do not handle, do not taste, do not touch!" (which all refer to things destined to perish with use) – in accordance with the commandments and teachings of men? These are matters which have, to be sure, the appearance of wisdom in self-made religion and self-abasement and severe treatment of the body, but are of no value against fleshly indulgence.* (Colossians 2:20-23)

4. Christian brothers and sisters are basically in the same "unit" when they are united in some group, such as a church, a Bible study, or a ministry team. According to *The United States Marine Essential Subjects* as quoted on page 38, our physical health can affect our unit's readiness and efficiency. This means we must encourage each other to maintain a decent physical fitness as well as discipline ourselves to be fit in other areas. How do the following verses tell us to help maintain physical fitness of our unit and even of other units?

   - *And the congregation of those who believed were of one heart and soul; and not one of them claimed that anything belonging to him was his own, but all things were common property to them. And with great power the apostles were giving testimony to the resurrection of the Lord Jesus, and abundant grace was upon them all, for all who were owners of land or houses would sell them and bring the proceeds of the sales and lay them at the apostles' feet, and they would be distributed to each as any had need.* (Acts 4:32-37)

   - *If a brother or sister is without clothing and in need of daily food, and one of you says to them, "Go in peace, be warmed and be filled," and yet you do not give them what is necessary for their body, what use is that?* (James 2:15-16)

---

[1] "Passion" in this verse is defined in Strong's as "inordinate affection" and "lust." Therefore, this word can be seen as an unhealthy or improper passion.

Now that you've looked at and considered several scriptures about physical health, I'd like to present some reasoning based on my own experiences and observations.

There are times we have little to no control over our physical injuries, illnesses, and challenges. Can we honestly say that these physical situations do not affect our spiritual, mental, or emotional condition? How many of us are at a point in our spiritual life where we could immediately rejoice and be strengthened by a new attack on our physical well-being? We may get to that point, but I know of very few who can immediately rejoice in their sufferings. It is the time between first experiencing a physical illness or injury and the moment of overcoming it (spiritually, mentally and/or physically) that I wish to focus on for a moment. I also include in that time period when one has a heightened awareness of or a struggle with a less-than-desired level of physical fitness because all these periods of time cause the Christian to be "a burden to the readiness and combat efficiency of his unit," as described in *The United States Marine Essential Subjects.*

During these times, people's focus is divided and too frequently they focus on their own issues instead of on what is going on around them. This makes them less ready for and/or less efficient at spiritual combat. I'm not saying this to condemn anyone or to say this makes someone who is physically out-of-shape unable to be a strong prayer warrior. I'm just saying that in my experience and through my own observations, I recognize that there are moments when, because of a physical battle we are currently engaged in, we are not as spiritually strong as we could be. On the other hand, sometimes a person in a physical battle becomes a great prayer warrior and wise saint, but this will be after they have conquered the normal human reaction to their physical struggle. Also, they, like all of us, will have their ups and downs as they continue to struggle with their issues and other events that come up in life.

What is my point concerning this time of weakness? First, the length and depth of this period is greatly affected by the Christian's perspective and choices. If my Marine Corps unit was getting ready to go out on a mission, I would do my best to be physically and mentally prepared for it so that my fellow-Marines could depend upon me. This would show that in my point of view, my comrades and the success of the mission are more important than my convenience or comfort. It would also show that I will make choices that will give the best results for those in my unit and for the mission.

Likewise, we believers should continue to learn all we can to prepare us for the battles that will come, and we should do what we can to maintain our health so that our bodies as well as our minds will be ready and efficient for our mission. We should try to be as fit as possible - physically, mentally, emotionally, and spiritually - because we don't know what attack will come our way next or what attack on a loved one will come. For example, I remember when my mother was very sick, I went to help her. At that time, my back was not doing well. This made it very painful and difficult for me to give her the help she needed, such as when I had to help her to the bathroom and she passed out in my arms. Only through great determination, experience in ignoring pain, and a tremendous amount of God's grace and power, was I able to get her where she needed to be without dropping her or screaming out in pain. This example shows that I had disciplined myself in areas for maintaining my fitness (even with a bad back), for making choices that would bring about success, and for relying on God. These are things we all need to incorporate into our lifestyle.

If there is anything we can do to improve our health, we should try to do so without fixating on it. In Matthew 26:36-46, when Jesus was in the Garden of Gethsemane and in prayer concerning what He was about to face, He had asked His disciples to keep watch. After a time of intense prayer by Himself, He went back to His disciples and found them asleep. He reprimanded them and included the comment, "the spirit is willing, but the flesh is weak." They were unable to keep their eyes open because they had not practiced what was needed in order to stay awake when the body insists on sleep. This made them "unfit" for that moment more than their sleepiness did. Whether we can make a difference with our physical challenges or not, it is most important to learn how to be overcomers whether these physical challenges remain with us or not. Part of overcoming is also to not be obsessed with our challenges. Do not let them take your mind or heart away from your spiritual growth. Learn to use the pain and the difficulties to drive you closer to your true source of real, permanent health – the Lord Jesus Christ. At the same time, be like Moses when he was too weak to hold up his arms and bring about victory for the Israelites. He allowed others to come alongside him and hold his arms up for him. (See Exodus 17:8-13.)

First Aid

In boot camp, everyone is taught basic first aid. I've included first aid within the section "health" because first aid is a way to regain health. In the Bible, the Christian is also given ways to maintain and to recover health. Some of these ways may seem difficult, impossible, or even cruel at times. However, we must remember two things: ❶ God will not ask or tell us to do something that He does not give us the tools and abilities to accomplish; ❷ often it isn't the finished goal, but the striving that matters most since it is God who does the accomplishing (Philippians. 1:6).

First aid is not just for oneself, but is to be used for our fellow-warriors (fellow-believers). Indeed, it is also used on those who are not on our side, for no medic worthy of the calling would refuse to render aid to "the enemy." Let's look at some Scriptures to see how all this relates in the spiritual.

1. Why and how are we called to care for our comrades in arms (brothers and sisters in Christ)?

- *"Two are better than one because they have a good return for their labor. For if either of them falls, the one will lift up his companion. But woe to the one who falls when there is not another to lift him up. Furthermore, if two lie down together they keep warm, but how can one be warm alone? And if one can overpower him who is alone, two can resist him. A cord of three strands is not quickly torn apart.* (Ecclesiastes 4:9-12)

- *For I long to see you so that I may impart some spiritual gift to you, that you may be established; that is, that I may be encouraged together with you while among you, each of us by the other's faith, both yours and mine.* (Romans 1:11-12)

- *And if one member suffers, all the members suffer with it; if one member is honored, all the members rejoice with it.* (1 Corinthians 12:26)

- *Brethren, even if anyone is caught in any trespass, you who are spiritual, restore such a one in a spirit of gentleness; each one looking to yourself, so that you too will not be tempted. Bear one another's burdens, and thereby fulfill the law of Christ.* (Galatians 6:1-2)

- *For God who has not destined us for wrath, but for obtaining salvation through our Lord Jesus Christ, who died for us, so that whether we are awake or asleep, we will live together with Him. Therefore, encourage one another and build up one another, just as you also are doing.* (1 Thessalonians 5:9-11)

- *We urge you, brethren, admonish the unruly, encourage the fainthearted, help the weak, be patient with everyone. See that no one repays another with evil for evil, but always seek after that which is good for one another and for all people.* (1 Thessalonians 5:14)

- *But encourage one another day after day, as long as it is still called "Today," so that none of you will be hardened by the deceitfulness of sin.* (Hebrews 3:13)

- *Therefore, confess your sins to one another, and pray for one another so that you may be healed. The effective prayer of a righteous man can accomplish much.* (James 5:16)

First aid, fundamentally, is about restoring a person to wholeness. These principles are equally as applicable for fellow believers as they are for those not yet aligned with our faith. Now, focusing on the practical aspects of first aid, we find four essential lifesaving steps in the following order: restore breathing, stop the bleeding, protect the wound, treat for shock. Let's explore each of these in more detail, discovering their significance in our quest for healing and wholeness.

Restore breathing. I have found various sources that generally say, a person can go without air for about 3 minutes. That's not very long. Therefore, getting them breathing is the highest priority. How does this relate to the spiritual?

Understand that "breath" and "spirit" are often the same word in Hebrew, so when you see "spirit" or "spirits," in the Old Testament verses below, you can often replace it with the word "breath" and vice versa.

2. Where does our physical breath come from?
   - *Then the LORD God formed man of the dust from the ground, and breathed into his nostrils the breath of life; and man became a living being.* (Genesis 2:7)
   - *The Spirit of God has made me, and the breath of the Almighty gives me life.* (Job 33:4)
   - *Thus says the Lord GOD to these bones, "Behold, I will cause breath to enter you that you may come to life. I will put sinews on you, make flesh grow back on you, cover you with skin and put breath in you that you may come alive; and you will know that I am the LORD."* (Ezekiel 37:5-6)
   - *But after the three and a half days, the breath of life from God came into them, and they stood on their feet; and great fear fell upon those who were watching them.* (Revelation 11:11)

3. How can we lose our physical breath?
   - *Behold, I, even I am bringing the flood of water upon the earth, to destroy all flesh in which is the breath of life, from under heaven; everything that is on the earth shall perish.* (Genesis 6:17)

- *Abraham breathed his last and died in a ripe old age, an old man and satisfied with life; and he was gathered to his people.* (Genesis 25:8)

- *Thus Joshua struck all the land, the hill country and the Negev and the lowland and the slopes and all their kings. He left no survivor, but he utterly destroyed all who breathed, just as the LORD, the God of Israel, had commanded.* (Joshua 10:40)

- *Now it came about after these things that the son of the woman, the mistress of the house, became sick; and his sickness was so severe that there was no breath left in him.* (1 Kings 17:17)

- *According to what I have seen, those who plow iniquity and those who sow trouble harvest it. By the breath of God they perish, and by the blast of His anger they come to an end.* (Job 4:8-9)

- *If He should determine to do so, if He should gather to Himself His spirit and His breath, all flesh would perish together, and man would return to dust.* Job 34:14-15

4. Where does our spiritual breath come from?

    - *And when He* [Jesus] *had said this, He breathed on them and said to them, "Receive the Holy Spirit."* (John 20:22)

    - *The God who made the world and all things in it, since He is Lord of heaven and earth, does not dwell in temples made with hands; nor is He served by human hands, as though He needed anything, since He Himself gives to all people life and breath and all things* (Acts 17:24-25)

- There is an imitator. Answer the following two questions after reading the associated scriptures.

    - Who gives breath to the image, and what is the purpose of giving it breath? *[Notice that the image has breath and can speak, but it is not alive.] Then I saw another beast coming up out of the earth; and he had two horns like a lamb and he spoke as a dragon. He exercises all the authority of the first beast in his presence. And he makes the earth and those who dwell in it to worship the first beast, whose fatal wound was healed. He performs great signs, so that he even makes fire come down out of heaven to the earth in the presence of men. And he deceives those who dwell on the earth because of the signs which it was given him to perform in the presence of the beast, telling those who dwell on the earth to make an image to the beast who had the wound of the sword and has come to life. And it was given to him to give breath to the image of the beast, so that the image of the beast would even speak and cause as many as do not worship the image of the beast to be killed.* (Revelation 13:11-15)

- Why would this action cause many to worship the Beast?

5. How can we lose our spiritual breath?

- *The idols of the nations are but silver and gold, the work of man's hands. They have mouths, but they do not speak; they have eyes, but they do not see; they have ears, but they do not hear, nor is there any breath at all in their mouths. Those who make them will be like them, yes, everyone who trusts in them.* (Psalm 135:15-18)

- *For whoever wishes to save his life will lose it; but whoever loses his life for My sake will find it.* (Matthew 16:25)

6. How can we restore our spiritual breath?

- *So it shall be when all of these things have come upon you, the blessing and the curse which I have set before you, and you call them to mind in all the nations where the LORD your God has banished you, and you return to the LORD your God and obey Him with all your heart and soul according to all that I command you today, you and your sons, then the LORD your God will restore you from captivity, and have compassion on you, and will gather you again from all the peoples where the LORD your God has scattered you.* (Deuteronomy 30:1-3)

- *Then he will pray to God, and He will accept him, that he may see His face with joy, and He may restore His righteousness to man. He will sing to men and say, "I have sinned and perverted what is right, and it is not proper for me. He has redeemed my soul from going to the pit, and my life shall see the light."* (Job 33:26-28)

- *Create in me a clean heart, O God, and renew a steadfast spirit within me. Do not cast me away from Your presence and do not take Your Holy Spirit from me. Restore to me the joy of Your salvation and sustain me with a willing spirit.* (Psalm 51:10-12)

Stop the Bleeding. After determining the injured is breathing, the next thing to do is to check for injuries. Wounds that are bleeding need to be treated in a specific sequence: First apply over a wound a dressing that does not slip and is sealed so no dirt gets in. If bleeding does not stop after applying the dressing, pressure may need to be applied to the wound without removing the dressing. The final step if the wound does not stop bleeding and is on an extremity would be to apply a tourniquet.

1. Let's first see what Scripture says about blood that is pertinent concerning first aid. Write what you learn about blood after each of these verses.

   - *He [the LORD] said, "What have you done? The voice of your brother's blood is crying to Me from the ground."* (Genesis 4:10)

   - *Whoever sheds man's blood, by man his blood shall be shed, for in the image of God He made man.* (Genesis 9:6)

   - *So Moses took the blood and sprinkled it on the people, and said, "Behold the blood of the covenant, which the LORD has made with you in accordance with all these words." (Exodus 24:8)*

   - *For the life of the flesh is in the blood…* (Leviticus 17:11a)

   - *So you shall not pollute the land in which you are; for blood pollutes the land and no expiation can be made for the land for the blood that is shed on it, except by the blood of him who shed it.* (Numbers 35:33)

   - *So you shall remove the guilt of innocent blood from your midst, when you do what is right in the eyes of the LORD.* (Deuteronomy 21:9)

   - *And He said to them, "This is My blood of the covenant, which is poured out for many."* (Mark 14:24)

- *So Jesus said to them, "Truly, truly, I say to you, unless you eat the flesh of the Son of Man and drink His blood, you have no life in yourselves."* (John 6:53)

- *Be on guard for yourselves and for all the flock, among which the Holy Spirit has made you overseers, to shepherd the church of God which He purchased with His own blood.* (Acts 20:28)

- *but if we walk in the Light as He Himself is in the Light, we have fellowship with one another, and the blood of Jesus His Son cleanses us from all sin.* (1 John 1:7)

So, we learn through these verses that blood is very important. It is necessary for life. Indeed, it is equal to life. Nobody can live without blood in them. When a life is taken, the blood cries out to God. This means that God values life and He knows when it has expired. In Matthew 10:29-31, Jesus states that He knows when a sparrow dies, and He cares even more about us than about sparrows. Then, when innocent blood is taken, His word tells us it pollutes the land, and the only way to cleanse the land is by rightly judging the one who shed it and by taking appropriate action according to God. We also learn that a covenant (a very serious treaty, contract, or agreement) can be confirmed, sealed, or signed with blood. When this is done, it is basically saying that the covenant makers are willing to die in order to fulfill the promises made in that contract. That is something every military person does when they swear the oath. They do not need to bleed when they swear the oath, but in order to keep it, they might need to and they are willing to.

When Jesus spoke of us drinking His blood, He was promising that His covenant with us to pay for our sins was serious enough to require a sacrifice – Himself. He was also stating that in order for us to receive all the benefits of that contract, we would need His life in us. We would need to fully partake of everything He was offering to us, which includes His life. I see this as being somewhat similar to the soldier in combat who throws himself upon a grenade in order to protect his comrades. His blood and body may be strewn all over, including onto those near him, but he hopes to absorb enough of the blast to save all but himself. Those who live because of his sacrifice will then be left with the thought of how to be worthy of his sacrifice, how to make the gift of his life for theirs to not be in vain.

What would it mean for us to bleed in a spiritual sense? It would mean that something is draining the life of Christ out of us. It could even mean that something is draining the life out of a church because a church is really a body of believers, not a building.

2. The Christian should regularly inspect himself or herself to be sure no wound is causing blood flow. Write what you learn from the following verses about searching for and discovering areas that could cause us to bleed.

- *Therefore, whoever eats the bread or drinks the cup of the Lord in an unworthy manner, shall be guilty of the body and the blood of the Lord. But a man must examine himself, and in so doing he is to eat of the bread and drink of the cup.* (1 Corinthians 11:27-28)

- *The heart is more deceitful than all else and is desperately sick; who can understand it? I, the LORD, search the heart, I test the mind, even to give each man according to His ways, according to the results of his deeds.* (Jeremiah 17:9-10)

- *As for you, my son Solomon, know the God of your father, and serve Him with a whole heart and a willing mind; for the LORD searches all hearts, and understands every intent of the thoughts. If you seek Him, He will let you find Him; but if you forsake Him, He will reject you forever.* (1 Chronicles 28:9)

- *O LORD, You have searched me and known me. You know when I sit down and when I rise up; You understand my thought from afar. You scrutinize my path and my lying down, and are intimately acquainted with all my ways.* (Psalm 139:1-3)

- *Search me, O God, and know my heart; try me and know my anxious thoughts; and see if there be any hurtful way in me, and lead me in the everlasting way.* (Psalm 139:23-24)

- *I, the LORD, search the heart, I test the mind, even to give to each man according to his ways, according to the results of his deeds.* (Jeremiah 17:10)

- *and He who searches the hearts knows what the mind of the Spirit is, because He intercedes for the saints according to the will of God.* (Romans 8:27)

So, we learn through the previous verses that God searches us thoroughly. He knows us better than we know ourselves. Therefore, it would be logical to go to Him to learn about ourselves rather than to only trust our own discernment. God is the One who can help us see what really motivates us, where our weaknesses are, and where we are "bleeding." And because He knows us so well, He also knows how to bring us into alignment with His will for us.

3. Once found, the "wound" needs to be treated correctly. Applying dressing to a bleeding wound is for the purpose of absorbing the flow until it coagulates. After it clots, healing can begin. If it continues to flow, pressure is needed to slow the flow to a speed that will allow the congealing. If even this does not work on an extremity, the flow must be cut off. Thus, a tourniquet would be needed. Here are some examples of how this process can be seen in the spiritual life of Christians. Notice that the tourniquet is the LAST resort. Observe also that there is always a desire to be able to remove the tourniquet and allow the full restoration of the "limb."

    - Self

        - Using dressing.

            - Who performs this healing treatment according to Psalm 147:2-3? *The LORD builds up Jerusalem; He gathers the outcasts of Israel. He heals the brokenhearted and binds up their wounds.*

            - How should we participate with the first aid process? *Do not be wise in your own eyes; fear the LORD and turn away from evil. It will be healing to your body and refreshment to your bones."* (Proverbs 3:7-8)

            How can we apply dressing to our own wounds? King David sinned greatly in the matter of Bathsheba and he was blind to his own "bleeding wound" (sin) until the prophet Nathan pointed it out to him. Because he truly loved God and wanted to serve him, this revelation by Nathan hit him hard and he was able to see through God's eyes the truth of what he had done. He was able to see his wound. Psalm 51 is a heart-felt poem describing his reaction. In the Psalm, David states that he is well aware of his sins and that he is utterly destitute without the Lord's compassionate cleansing and renewing of "a steadfast spirit within" him. He acknowledges that nothing he does can repair what his sin has destroyed – his relationship with God. He repents for his sin and expresses his desire to act in line with his repentance, requesting that God, Himself, restore a clean heart within him so that their relationship can also be restored and his actions from the restored relationship will once again bring glory to God and draw others to the Lord.

        - Using pressure. God often uses the pressure of difficulties to cause us to see our "wounds" and needs so that we will turn to Him for healing and health. What do you learn about this process from the following verses?

- *O LORD, rebuke me not in Your wrath, and chasten me not in Your burning anger. For Your arrows have sunk deep into me, Your hand has pressed down on me. There is no soundness in my flesh because of Your indignation; there is no health in my bones because of my sin.* (Psalm 38:1)

- *Then they cried out to the LORD in their trouble; He saved them out of their distresses. He sent His word and healed them, and delivered them from their destructions.* (Psalm 107:19-20)

- *Come, let us return to the LORD. For He has torn us, but He will heal us; He has wounded us, but He will bandage us.* (Hosea 6:1)

- Write about a time when God used difficulties to draw you nearer to Him or even to turn you from something to Him.

- Using a tourniquet
  - *If your hand causes you to stumble, cut it off; it is better for you to enter life crippled, than, having your two hands, to go into hell, into the unquenchable fire, where their worm does not die, and the fire is not quenched. If your foot causes you to stumble, cut it off; it is better for you to enter life lame, than, having your two feet, to be cast into hell, where their worm does not die, and the fire is not quenched. If your eye causes you to stumble, throw it out; it is better for you to enter the kingdom of God with one eye, than, having two eyes, to be cast into hell, where their worm does not die, and the fire is not quenched.* (Mark 9:43-48)
  - In Mark 9, these are all examples of "using a tourniquet" for the purpose of showing the degree of commitment we should have concerning doing what is right in God's eyes. It is not an encouragement to actually cut off body parts.
  - An example of what is encouraged here would be an addiction I dealt with concerning video games. At that time, video game centers were very popular. I found I could not leave one without having spent every cent I had with me. I was unable to control my spending once I entered, no matter how determined I was nor how many "rules" I put upon myself prior to entering. Therefore, I decided to "cut off" this type of entertainment from my life. I never entered another video

game center until many decades later. Fortunately, when I did finally go into such a place, I had little to no desire to even play any of the games, though the friends I was with played for quite a while. When I did play a game, I was fine playing just once, then stopping. I was over the addiction, praise God!

- To make this more personal, write an example of the use of a spiritual tourniquet of which you are familiar.

- Within the church.
  - Matthew 18:15-17 has examples of the 3 different treatments to be attempted for the bleeding of a brother. Can you find them?

    "If your brother sins, go and show him his fault in private; if he listens to you, you have won your brother. But if he does not listen to you, take one or two more with you, so that by the mouth of two or three witnesses every fact may be confirmed. If he refuses to listen to them, tell it to the church; and if he refuses to listen even to the church, let him be to you as a Gentile and a tax collector."

An example of God's treatment of an extremely threatening wound, like one with gangrene, can be found in Revelation 2:20-23.

- *But I have this against you, that you tolerate the woman Jezebel, who calls herself a prophetess, and she teaches and leads My bond-servants astray so that they commit acts of immorality and eat things sacrificed to idols. I gave her time to repent, and she does not want to repent of her immorality. Behold, I will throw her on a bed of sickness, and those who commit adultery with her into great tribulation, unless they repent of her deeds. And I will kill her children with pestilence, and all the churches will know that I am He who searches the minds and hearts; and I will give to each one of you according to your deeds.* (Revelation 2:20-23)
  - We don't read specifically what God's "dressing" was, but we can see that it was presented along with time to repent. However, Jezebel refused it. What is the "pressure" God will apply, and what is the "tourniquet?"

- What is the way for those involved with or following Jezebel to avoid needing to receive increasingly serious treatments?

- The Corinthian church had an issue where Paul suggested using a tourniquet because this was an extremely dangerous spiritual situation. Why was it so dangerous? What is the tourniquet Paul recommends?

  It is actually reported that there is immorality among you, and immorality of such a kind as does not exist even among the Gentiles, that someone has his father's wife. You have become arrogant and have not mourned instead, so that the one who had done this deed would be removed from your midst. For I, on my part, though absent in body but present in spirit, have already judged him who has so committed this, as though I were present. In the name of our Lord Jesus, when you are assembled, and I with you in spirit, with the power of our Lord Jesus, I have decided to deliver such a one to Satan for the destruction of his flesh, so that his spirit may be saved in the day of the Lord Jesus. Your boasting is not good. Do you not know that a little leaven leavens the whole lump of dough? (1 Corinthians 5:1-6)

- In 2 Corinthians 2:4-11, Paul addresses this situation again after the church took action. What insights do you glean from these scriptures concerning the reasons Paul had determined such a drastic measure was needed? What is the result of these harsh measures?

  For out of much affliction and anguish of heart I wrote to you with many tears; not so that you would be made sorrowful, but that you might know the love which I have especially for you. But if any has caused sorrow, he has caused sorrow not to me, but in some degree – in order not to say too much – to all of you. Sufficient for such a one is this punishment which was inflicted by the majority, so that on the contrary you should rather forgive and comfort him, otherwise such a one might be overwhelmed by excessive sorrow. Wherefore I urge you to reaffirm your love for him. For to this end also I wrote, so that I might put you to the test, whether you are obedient in all things. But one whom you forgive anything, I forgive also; for indeed what I have forgiven, if I have forgiven anything, I did it for your sakes in the presence of Christ, so that no advantage would be taken of us by Satan, for we are not ignorant of his schemes.

This situation in the Corinthian church can also be used as a spiritual equivalent to the final two things to do in a bleeding situation: protect the wound and treat for shock. After the church obeyed Paul and put this man out of their congregation, he was extremely sorrowful. If he had remained in this condition outside of the church family, things could have gotten a lot worse for him as well as for the church. Infection in the body of believers could have crept in with results of things like animosity towards the leaders, or possibly a self-righteousness legalism springing up. So, Paul was admonishing them to "protect the wound." Additionally, the young man who had sinned could have turned back to the world, believing Christianity was too difficult or feeling they were hypocritical, or he could have become so self-condemning that he would do harm to himself. In forgiving him and bringing him back into the fold, the church was "treating him for shock."

NBC (Nuclear, Biological, & Chemical) Defense

Though the USA is capable of NBC attacks, we should not be using them because they kill indiscriminately, injuring many innocents, and because they have long-term effects on the environment and on society. However, it is possible that our enemies may use them. Therefore, it is important that we learn how to defend against such attacks as best as possible.

For each of these types of attacks, recruits learn how to recognize what type of attack is being waged against them, the best location and position to be in for survival, and how to use the tools at their disposal. For example, if it is a chemical attack, we signal to our comrades the kind of attack, then don our masks and get under our ponchos in such a way that no chemical could leach in to attack us.

It is important to recognize the kind of attack we are experiencing so that the proper defensive strategy can be utilized. What kind of attacks can be waged against the Christian? There are basically two categories: Physical and Spiritual. Let's look at scriptures which tell us how to respond to such attacks, how to properly defend against them, and what tools and weapons are at our disposal.

1. There are two types of physical attacks which the believers need to deal with, though they are not confined to just Christians: persecution and sickness. There are also spiritual attacks. How should we respond to these kinds of attacks?
    a. Physical Persecution. Because human persecutors are not the Christian's true enemy, it should not be our preferred choice to fight back using conventional styles of warfare. Persecution can be viewed like a chemical attack and sometimes like a nuclear attack. We need to find ways to protect ourselves from and/or to diffuse the situation, but it is useless, often even dangerous, for us to attack the nuclear bomb or the chemical (the persecutors) because it isn't the true source of the attack and only produces hazardous materials (more hate in both the believer and the non-believer). Nevertheless, I will admit that there are situations where the persecutors will not be dissuaded by any other means than using physical force. Just remember that becoming physical is a last resort.

From the following verses, what do you learn about persecution of Believers and the position (attitude) they should take?

- *"Blessed are those who have been persecuted for the sake of righteousness, for theirs is the kingdom of heaven. Blessed are you when people insult you and persecute you and falsely say all kinds of evil against you because of Me. Rejoice and be glad, for your reward in heaven is great; for in the same way they persecuted the prophets who were before you."* (Matthew 5:10-12)

- *"But I say to you, do not resist an evil person; but whoever slaps you on your right cheek, turn the other to him also." "But I say to you, love your enemies and pray for those who persecute you,"* (Matthew 5:39, 44)

- *"But whenever they persecute you in one city, flee to the next; for I say to you, you will not finish going through the cities of Israel until the Son of Man comes."* (Matthew 10:23)

- *"But before all these things, they will lay their hands on you and will persecute you, delivering you to the synagogues and prisons, bringing you before kings and governors for My name's sake."* (Luke 21:12)

- *"Remember the word that I said to you, 'A slave is not greater than his master.' If they persecuted Me, they will also persecute you; if they kept My word, they will keep yours also."* (John 15:20)

- *Bless those who persecute you; bless and do not curse.* (Romans 12:14)

- *And we toil, working with our own hands; when we are reviled, we bless; when we are persecuted, we endure; when we are slandered, we try to conciliate; we have become as the scum of the world, the dregs of all things, even until now. I do not write these things to shame you, but to admonish you as my beloved children.* (1 Corinthians 4:12-14)

- *Indeed, all who desire to live godly in Christ Jesus will be persecuted.* (2 Timothy 3:12)

The previous verses help us see that persecution is to be expected for followers of Jesus. However, we are not to handle it as the world would, for those who persecute us are not our true enemy. They are only a tool in his hands.

b. Physical Sickness. Sickness is not our enemy either. It is just a tool by which the enemy of our souls wants to bring us down, but which God can use to build us up. This could be like a biological attack. At this point in time, no disease has been fully eradicated (though it may seem so); therefore, we can see that attacking the sickness itself is practically useless. It has often been "attacked," but not wiped out. Unless it is our medical field of expertise, our reaction instead should be to try to position ourselves to be less likely to succumb, with the understanding that we may not always be successful. Therefore, we need to be able to treat as well as prevent sicknesses, and perhaps develop immunity because we have survived. Ultimately, it is all in God's hands.

From the following verses, what do you learn about sickness? What do you discover about the beliefs and reactions to it which the Christian should take?

- *But when Jesus heard this, He said, "It is not those who are healthy who need a physician, but those who are sick."* (Matthew 9:12)

- *Jesus was going through all the cities and villages, teaching in their synagogues and proclaiming the gospel of the kingdom, and healing every kind of disease and every kind of sickness.* (Matthew 9:35)

- *Jesus summoned His twelve disciples and gave them authority over unclean spirits, to cast them out, and to heal every kind of disease and every kind of sickness.... "Heal the sick, raise the dead, cleanse the lepers, cast out demons. Freely you have received, freely give."* (Matthew 10:1, 8)

- *Then Peter, filled with the Holy Spirit, said to them, "Rulers and elders of the people, if we are on trial for a benefit done to a sick man, as to how this man has been made well, let it be known to all of you and to all the people of Israel, that by the name of Jesus Christ the Nazarene, whom you crucified, whom God raised from the dead – by this name this man stands here before you in good health."* (Acts 4:8-10)

- *Erastus remained at Corinth, but Trophimus I left sick at Miletus.* (2 Timothy 4:20) [Paul, who had performed miracles in Jesus's name, wrote this.]

- *Is anyone among you sick? Then he must call for the elders of the church and they are to pray over him, anointing him with oil in the name of the Lord; and the prayer offered in faith will restore the one who is sick, and the Lord will raise him up, and if he has committed sins, they will be forgiven him. Therefore, confess your sins to one another, and pray for one another so that you may be healed. The effective prayer of a righteous man can accomplish much.* (James 5:14-16)

- *"Behold, I will throw her on a bed of sickness, and those who commit adultery with her into great tribulation, unless they repent of her deeds. And I will kill her children with pestilence, and all the churches will know that I am He who searches the minds and hearts; and I will give to each one of you according to your deeds."* (Revelation 2:22-23)

c. Spiritual Attacks. Most of the time, spiritual attacks are subversive and subtle, and often come from within one's mind. The mind is the main battlefield for spiritual warfare. However, there are times we are attacked very openly from the spiritual realm, and sometimes these can manifest in the physical world.

   What do you learn from the following verses concerning spiritual fighting?

   - *Now it shall be, if you diligently obey the LORD your God, being careful to do all His commandments which I command you today, the LORD your God will set you high above all the nations of the earth....The LORD shall cause your enemies who rise up against you to be defeated before you; they will come out against you one way and will flee before you seven ways.* (Deuteronomy 28:1, 7)

- *"No weapon that is formed against you will prosper; and every tongue that accuses you in judgment you will condemn. This is the heritage of the servants of the LORD, and their vindication is from Me," declares the LORD.* (Isaiah 54:17)

- *Behold, I have given you authority to tread on serpents and scorpions, and over all the power of the enemy, and nothing will injure you.* (Luke 10:19)

- *For our struggle is not against flesh and blood, but against the rulers, against the powers, against the world forces of this darkness, against the spiritual forces of wickedness in the heavenly places.* (Ephesians 6:12)

- *Submit therefore to God. Resist the devil and he will flee from you.* (James 4:7)

2. A well-planned defense includes the best location and position to be in based on the kind of attack. For the Christian, this means where your mind is focused, what your mindset is, and what you have done to prepare for battle. Once you have prepared your defense, then you are ready for the attack. Both physical and spiritual battles require a well-planned defense.

    a. Physical. What do you learn from the following verses about attitudes and any preparatory actions that should be done for physical attacks?

        1) Persecution

        - *You shall not take vengeance, nor bear any grudge against the sons of your people, but you shall love your neighbor as yourself; I am the LORD.* (Leviticus 19:18)

        - *Vengeance is Mine, and retribution, in due time their foot will slip; for the day of their calamity is near, and the impending things are hastening upon them.* (Deuteronomy 32:35)

- *David said to all his servants who were with him at Jerusalem, "Arise and let us flee, for otherwise none of us will escape from Absalom. Go in haste, or he will overtake us quickly and bring down calamity on us and strike the city with the edge of the sword."* (2 Samuel 15:14)

- *Then some came and reported to Jehoshaphat, saying, "A great multitude is coming against you from beyond the sea, out of Aram and behold, they are in Hazazon-tamar (that is Engedi)." Jehoshaphat was afraid and turned his attention to seek the LORD, and proclaimed a fast throughout all Judah.* (2 Chronicles 20:2-3)

- *In the morning, O LORD, You will hear my voice; in the morning I will order my prayer to You and eagerly watch.* (Psalm 5:3)

- *The LORD lives, and blessed be my rock; and exalted be the God of my salvation, the God who executes vengeance for me, and subdues peoples under me. He delivers me from my enemies; surely You lift me above those who rise up against me; You rescue me from the violent man.* (Psalm 18:46-48)

- *You will not be afraid of the terror by night, or of the arrow that flies by day.* (Psalm 91:5)

- *Now when they had gone, behold, an angel of the Lord appeared to Joseph in a dream and said, "Get up! Take the Child and His mother and flee to Egypt, and remain there until I tell you; for Herod is going to search for the Child to destroy Him."* (Matthew 2:13)

- *Behold, I send you out as sheep in the midst of wolves; so be shrewd as serpents and innocent as doves. But beware of men, for they will hand you over to the courts and scourge you in their synagogues; and you will even be brought before governors and kings for My sake, as a testimony to them and to the Gentiles. But when they hand you over, do not worry about how or what you are to say; for it will be given you in that hour what you are to say. For it is not you who speak, but it is the*

Spirit of your Father who speaks in you. Brother will betray brother to death, and a father his child; and children will rise up against parents and cause them to be put to death. You will be hated by all because of My name, but it is the one who has endured to the end who will be saved. But whenever they persecute you in one city, flee to the next; for truly I say to you, you will not finish going through the cities of Israel until the Son of Man comes. (Matthew 10:16-23)

- *For where two or three have gathered together in My name, I am there in their midst.* (Matthew 18:20)

- *Never take your own revenge, beloved, but leave room for the wrath of God, for it is written, "Vengeance is Mine, I will repay," says the Lord.* (Romans 12:19)

- *And let us consider how to stimulate one another to love and good deeds, not forsaking our own assembling together, as is the habit of some, but encouraging one another; and all the more as you see the day drawing near.* (Hebrews 10:24-25)

- *So then, let us not sleep as others do, but let us be alert and sober.* (1 Thessalonians 5:6)

2) Sickness (*I used some of the same verses in both persecution and in sickness. Can you see how they apply to each type of attack?*)

   - *I will say to the LORD, "My refuge and my fortress, My God, in whom I trust!" For it is He who delivers you from the snare of the trapper and from the deadly pestilence…You will not be afraid of the terror by night, or of the arrow that flies by day; of the pestilence that stalks in darkness, or of the destruction that lays waste at noon.* (Psalm 91:2-3, 5-6)

- *Wherever He entered villages, or cities, or countryside, they were laying the sick in the market places, and imploring Him that they might just touch the fringe of His cloak; and as many as touched it were being cured.* (Mark 6:56)

- *Because of the surpassing greatness of the revelations, for this reason, to keep me from exalting myself, there was given me a thorn in the flesh, a messenger of Satan to torment me – to keep me from exalting myself! Concerning this I implored the Lord three times that it might leave me. And He has said to me, "My grace is sufficient for you, for power is perfected in weakness." Most gladly, therefore, I will rather boast about my weaknesses, so that the power of Christ may dwell in me. Therefore, I am well content with weaknesses, with insults, with distresses, with persecutions, with difficulties, for Christ's sake; for when I am weak, then I am strong.* (2 Corinthians 12:7)

b. Spiritual. What do you learn from the following verses about attitudes and any preparatory actions that should be done for spiritual attacks?

- *Jehoshaphat was afraid and turned his attention to seek the LORD, and proclaimed a fast throughout all Judah. So Judah gathered together to seek help from the LORD; they even came from all the cities of Judah to seek the LORD.* (2 Chronicles 20:3-4)

- *In the morning, O LORD, You will hear my voice; in the morning I will order my prayer to You and eagerly watch.* (Psalm 5:3)

- *He who dwells in the shelter of the Most High will abide in the shadow of the Almighty.* (Psalm 91:1)

- *After He had sent the crowds away, He went up on the mountain by Himself to pray; and when it was evening, He was there alone.* (Matthew 14:23)

- *For where two or three have gathered together in My name, I am there in their midst.* (Matthew 18:20)

- *Keep watching and praying that you may not come into temptation; the spirit is willing, but the flesh is weak.* (Mark 14:38)

- *But Jesus Himself would often slip away to the wilderness and pray.* (Luke 5:16)

- *For though we walk in the flesh, we do not war according to the flesh, for the weapons of our warfare are not of the flesh, but divinely powerful for the destruction of fortresses. We are destroying speculations and every lofty thing raised up against the knowledge of God, and we are taking every thought captive to the obedience of Christ.* (2 Corinthians 10:3-5)

- *Be of sober spirit, be on the alert. Your adversary, the devil, prowls around like a roaring lion, seeking someone to devour.* (1 Peter 5:8)

3. What tools or weapons do we have and how do we use them?

   a. We have clothing (to be discussed later).

   b. We have some special tools. What are they and how are they used? Some of these verses are not clear in how a tool/weapon is to be used; therefore, you may need to spend some time meditating on how it is beneficial. I've given you an example on how to do this with Psalm 84:11.

      - *For the LORD God is a sun and shield; the LORD gives grace and glory; no good thing does He withhold from those who walk uprightly.* (Psalm 84:11)
        - First, what are the tools/weapons in this verse? The LORD is a sun and a shield. What do these do for us? A sun enables us to see clearly, it gives us energy, and it warms us. A shield protects us. Therefore, the LORD

protects us, enables us to see clearly, energizes us, and warms or comforts us.

- Next, the LORD gives grace and glory. How do these benefit us? Grace is for when we make mistakes or we don't succeed to the degree that would be best. He forgives us and encourages us, giving us what is needed to press on and to grow. His glory that He bestows on us enables others to see Him through us (even in our mistakes if we accept His grace) and possibly even exalts us in the process.

- Finally, God provides all good things necessary to help us as we do our best to walk uprightly with Him.

- Though some of these tools can be used offensively, they can also be used defensively as follows. The sun gives us clear sight so we won't believe lies of the enemy and we'll be able to prepare for his attacks. The shield gives us protection against the enemy's arrows and other weapons with which he seeks to wear us down or wound us so we'll be less effective. The energy from the sunlight helps keep us pressing on, and the warmth comforts us when we have our moments of exhaustion or pain.

In the following verses, state what the tools are and how they can be used.

- *When they began singing and praising, the LORD set ambushes against the sons of Ammon, Moab and Mount Seir, who had come against Judah; so they were routed.* (2 Chronicles 20:22)

- *He will cover you with His pinions, and under His wings you may seek refuge; His faithfulness is a shield and bulwark.* (Psalm 91:5)

- *A gentle answer turns away wrath, but a harsh word stirs up anger.* (Proverbs 15:1)

- *But you, when you pray, go into your inner room, close your door and pray to your Father who is in secret, and your Father who sees what is done in secret will reward you.* (Matthew 6:6)

- *Again I say to you that if two of you agree on earth about anything that they may ask, it shall be done for them by My Father who is in heaven. For where two or three have gathered together in My name, I am there in their midst.* (Matthew 18:19-20)

- *Behold, I have given you authority to tread on serpents and scorpions, and over all the power of the enemy, and nothing will injure you. Nevertheless do not rejoice in this, that the spirits are subject to you, but rejoice that your names are recorded in heaven.* (Luke 10:19-20)

- *But about midnight Paul and Silas were praying and singing hymns of praise to God, and the prisoners were listening to them; and suddenly there came a great earthquake, so that the foundations of the prison house were shaken; and immediately all the doors were opened and everyone's chains were unfastened.* (Acts 16:25-26)

- *So faith comes from hearing, and hearing by the word of Christ.* (Romans 10:17)

- *With all prayer and petition pray at all times in the Spirit and with this in view, be on the alert with all perseverance and petition for all the saints.* (Ephesians 6:18)

- *For the word of God is living and active and sharper than any two-edged sword, and piercing as far as the division of soul and spirit, of both joints and marrow, and able to judge the thoughts and intentions of the heart.* (Hebrews 4:12)

- *Knowing that you were not redeemed with perishable things like silver or gold from your futile way of life inherited from your forefathers, but with the precious blood, as of a lamb unblemished and spotless, the blood of Christ.* (1 Peter 1:17-18)

- *And they overcame him because of the blood of the Lamb and because of the word of their testimony, and they did not love their life even when faced with death.* (Revelation 12:11)

I have only skimmed the surface of the vast array of tools and weapons that God has given us for our spiritual warfare. Some are easy to recognize as weapons, some are not. For example, salvation is an awesome tool, for it puts us on the winning side, and it instills in us a strength, courage, and peace beyond all compare if we allow the fullness of our unworthiness combined with His grace upon us to permeate our minds and hearts. It enables soldiers to fully know that no matter the circumstances we see, we win (individually as well as collectively).

Chapter 4: Uniform Regulations and Clothing

There are basically two kinds of regular uniforms – the "dress" uniform and the "utility," "camouflage," or "combat" uniform. The uniform for each branch of service is different, but for the Christian, there is only one "branch," in spite of the Marine Corps anthem's last stanza – "If the Army and the Navy ever look on heaven's scene, they will find the streets are guarded by United States Marines." The general regulations for military uniforms relate to the correct uniform for the season and the occasion. They also relate to the care and presentation of the uniform and all the additions to the uniform known as rank insignia, ribbons, and metals.

1. What does Scripture say about our everyday presentation (everyday uniform)? How should we present ourselves to others and why or how are we able to do so?

    - *"But this is the covenant which I will make with the house of Israel after those days," declares the LORD, "I will put My law within them and on their heart I will write it; and I will be their God, and they shall be My people."* (Jeremiah 31:33)

    - *The LORD said to him, "Go through the midst of the city, even through the midst of Jerusalem, and put a mark on the foreheads of the men who sigh and groan over all the abominations which are being committed in its midst."* (Ezekiel 9:4)

    - *And I will give them one heart, and put a new spirit within them. And I will take the heart of stone out of their flesh and give them a heart of flesh."* (Ezekiel 11:19)

    - *Whenever you fast, do not put on a gloomy face as the hypocrites do, for they neglect their appearance so that they will be noticed by men when they are fasting. Truly I say to you, they have their reward in full.* (Matthew 6:16)

    - *And put on the new self, which in the likeness of God has been created in righteousness and holiness of the truth.* (Ephesians 4:24)

    - *Do not lie to one another, since you laid aside the old self with its evil practices, and have put on the new self who is being renewed to a true knowledge according to the image of the One who created him…So, as those who have been chosen of God holy and beloved, put on a heart of compassion, kindness, humility, gentleness and patience; bearing with one another, and forgiving each other, whoever has a complaint against anyone; just as the Lord forgave you, so also should you. Beyond all these things put on love, which is the perfect bond of unity.* (Colossians 3:9-10, 12-14)

- *Your adornment must not be merely external – braiding the hair, and wearing gold jewelry, or putting on dresses; but let it be the hidden person of the heart, with the imperishable quality of a gentle and quiet spirit, which is precious in the sight of God.* (1 Peter 3:3-4)

The other articles of clothing to know how to correctly wear relate to actual warfare. Some of these will be for the general warrior while others will be for specific jobs in battle conditions. I will cover these in my book on combat.

Chapter 5: Drill

I would be remiss if I did not mention drill (marching) since a large portion of the diary which I kept throughout boot camp had pages and pages in which all I wrote was "Classes and Drill." From the very first day of basic training, recruits march in formation from one location to another. It begins with being told how to stand, where to stand, and how much room between each person there should be. After that, learning how to turn in a military fashion is taught. Then, the unit learns how to "walk" (march) in rhythm together using a cadence which is called by a drill instructor. Often, the drill instructor, also called a DI, will chant something and the recruits are to repeat the chant.

Why is drill so important that it is used in every branch of the service? It isn't just because it looks great, especially when it is made more intricate. The main purpose for drill is to build unity. A group of strangers learn to move together as a unit. Unity is actually a main purpose for a great many "exercises" which recruits go through in boot camp.

There are divisions of groups of military personnel, the smallest being a Squad, then a Platoon. There are larger groups than these, but I won't go into that here since during most of boot camp, interaction in larger groups than these are very infrequent. Recruits are divided into platoons on their first day. These are the people they will bunk and shower with during basic training. They will often join one or more platoons for drill, classes, exercise, and chow, yet they will still be pretty much confined to their own platoon, often competing against other platoons.

In boot camp, individuality is not strongly encouraged, yet drill instructors keep an eye out for those who show leadership qualities. The drill instructors may place some of those people into positions of leadership. One such position is the Guide. This person carries the platoon flag, known as a guidon (pronounced "guide on"), which can be seen all the way to the back of the platoon. When marching, this position is very important for keeping the rest of the platoon moving in the correct direction at the right time. For example, when the drill instructor commands, "Column left!" the Guide makes an immediate 90 degree turn to the left. The person behind the Guide will do likewise once he/she reaches the place the Guide turned left. The rest in that line will do likewise. Usually, platoons have 4 lines of people. Each line could not make an exact 90 degree turn or they would run into each other. Therefore, the other columns make 2 turns, each with angles of 45 degrees, but with a different number of steps between each turn. Sound confusing? Here's a picture to help you visualize (G=Guide, X=leader of each column):

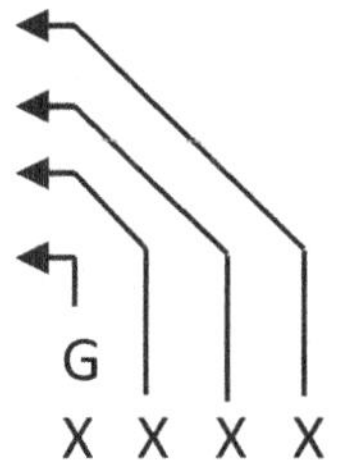

Figure 1: Column Left Marching Diagram

Remember that I said that drill was to help create unity? That isn't the only reason for drill. It is also to teach how to follow orders quickly. I was the Guide through most of my boot camp experience, so it was very important that I followed the commands of the DI because the rest of the platoon depended on me. At the time we were training to be Marines, it was apparent that we were also training a new drill instructor. She was not yet very good at judging when to command us to turn left or right (often missing the street by a few footsteps), so we learned to slide into the positions we knew she wanted. In other words, we didn't obey her orders immediately, but would take a step or more beyond the timing of her command before we would turn. One time, though, her command, "Column left!" came about 5-6 steps before the street. In an instant, I realized that I could not take the needed steps to the street without it being apparent I was ignoring her order. Therefore, I did an immediate left turn and marched straight towards a dumpster. However, I was alone in my actions. (So much for unity.) The rest of the platoon had done as usual and kept going until they came to the street, then they turned left. I well remember that I was wondering two things at the same time. First, I wondered if I was going to run into the dumpster before the drill instructor called out, "Column, halt!" My other thought was, "You're in for big trouble now, Cristi," because I was all by my little lonesome. What actually happened, though, was that the platoon got in trouble. I had followed my orders to the letter and without hesitation. Nobody else had. They just anticipated what the DI wanted and they did that. A Christian lesson in this is that we need to trust the One who gives the order more than we should trust our own understanding. As it states in Proverbs 3:5, *"Trust in the LORD with all your heart and do not lean on your own understanding."*

In our Christian life, we often go through "drills" in which we do the same things over and over again. Sometimes we'll understand what the purpose is, and sometimes we'll have absolutely no clue where we're heading. God uses these things to teach us unity with our Drill Instructor, the Holy Spirit, as we learn to recognize His voice and His gentle guidance. We also develop unity with our fellow believers as they also go through similar experiences. We, in this way, are able to encourage one another, sometimes following in another's footsteps who has successfully made it through a difficulty similar to what we are currently dealing with. We also develop leadership qualities, even if we do not consider ourselves leaders. Being examples for others, even if we never say a word, is a leadership skill.

1. What are some of the "drills" we go through and how should we respond? Write what we do or experience over and over which are shown in the following verses. If you see a suggested response to "the drill," write it, too.

    - *You shall keep My statutes and practice them; I am the LORD who sanctifies you.* (Leviticus 20:8)

    - *For Ezra had set his heart to study the law of the LORD and to practice it, and to teach his statutes and ordinances in Israel.* (Ezra 7:10)

- *How blessed are those who keep justice, who practice righteousness at all times!* (Psalm 106:3)

- *For if you truly amend your ways and your deeds, if you truly practice justice between a man and his neighbor, if you do not oppress the alien, the orphan, or the widow, and do not shed innocent blood in this place, nor walk after other gods to your own ruin, then I will let you dwell in this place, in the land that I gave to your fathers forever and ever.* (Jeremiah 7:5-7)

    Consider each of the things Jeremiah 7 says NOT to do, and put them into your culture. For example, "oppress the alien" makes me think of a migrant worker being pressured to pick more crops than physically possible for that person, then being punished in some way (like docking them some money) for not achieving that goal.

- *Thus has the LORD of hosts said, "Dispense true justice and practice kindness and compassion each to his brother; and do not oppress the widow or the orphan the stranger or the poor; and do not devise evil in your hearts against one another."* (Zechariah 7:9-10)

- *But he who practices the truth comes to the Light, so that his deeds may be manifested as having been wrought in God.* (John 3:21) What does it look like to practice the truth?

- *No temptation has overtaken you but such as is common to man; and God is faithful, who will not allow you to be tempted beyond what you are able, but with the temptation will provide the way of escape also, so that you will be able to endure it.* (1 Corinthians 10:13)

- *Now as to the love of the brethren, you have no need for anyone to write to you, for you yourselves are taught by God to love one another; for indeed you do practice it toward all the brethren who are in all Macedonia. But we urge you, brethren, to excel still more.* (1 Thessalonians 4:9-10)

- *The things you have learned and received and heard and seen in me, practice these things, and the God of peace will be with you.* (Philippians 4:9)

- *Consider it all joy, my brethren, when you encounter various trials, knowing that the testing of your faith produces endurance. And let endurance have its perfect result, so that you may be perfect and complete, lacking in nothing.* (James 1:2-4)

- *Beloved, do not be surprised at the fiery ordeal among you, which comes upon you for your testing, as though some strange thing were happening to you; but to the degree that you share the sufferings of Christ, keep on rejoicing, so that also at the revelation of His glory you may rejoice with exultation.* (1 Peter 4:12:13)

- *Little children, make sure no one deceives you; the one who practices righteousness is righteous, just as He is righteous;* (1 John 3:7)

2. How should we react when we mess up during drill (when we sin)?

- *So it shall be when he becomes guilty in one of these, that he shall confess that in which he has sinned.* (Leviticus 5:5)

- *Then he shall confess his sins which he has committed, and he shall make restitution in full for his wrong and add to it one-fifth of it, and give it to him whom he has wronged.* (Numbers 5:7)

- *I acknowledged my sin to You, and my iniquity I did not hide; I said, "I will confess my transgressions to the LORD;" and You forgave the guilt of my sin. Selah.* (Psalm 32:5)

- *He who conceals his transgressions will not prosper, but he who confesses and forsakes them will find compassion.* (Proverbs 28:13)

- *If we confess our sins, He is faithful and righteous to forgive us our sins and to cleanse us from all unrighteousness.* (1 John 1:9)

Chapter 6: Inspections

Inspections begin almost immediately upon entering boot camp, and formal inspections begin a little while later. The informal inspections are done as the drill instructors observe and make judgments about the recruit's attitude, behaviors, physical fitness, intelligence, reactions, appearance, and other miscellaneous things. These inspections, both formal and informal, will continue for as long as the person is in the military, though the inspectors will change and their place of work will also be included in inspections.

Christians are always undergoing inspections by non-believers and by other Christians. Additionally, if you are anything like me, you are also frequently or constantly inspecting yourself. There is nothing wrong with this as long as it is done with the right attitude by both the inspecting person as well as the person being inspected. However, even if the person in charge of the inspection doesn't have the right attitude, the person being inspected truly answers only to God, so his/her attitude must remain in line with Scripture (no matter who the inspector is).

1. What do the following verses tell us about our judge (inspector)?

- *May the LORD judge between you and me, and may the LORD avenge me on you; but my hand shall not be against you.* (1 Samuel 24:12)

- *Judge me, O LORD my God, according to Your righteousness, and do not let them rejoice over me.* (Psalm 35:24)

- *Say among the nations, "The LORD reigns; Indeed, the world is firmly established, it will not be moved; He will judge the peoples with equity.* (Psalm 96:10)

- *Search me, O God, and know my heart; try me and know my anxious thoughts; and see if there be any hurtful way in me, and lead me in the everlasting way.* (Psalm 139:23-24)

- *And He will delight in the fear of the LORD, and He will not judge by what His eyes see, nor make a decision by what His ears hear; but with righteousness He will judge the poor, and decide with fairness for the afflicted of the earth; and He will strike the earth with the rod of His mouth, and with the breath of His lips He will slay the wicked.* (Isaiah 11:3-4)

- *"For not even the Father judges anyone, but He has given all judgment to the Son." "I can do nothing on My own initiative. As I hear, I judge; and my judgment is just, because I do not seek My own will, but the will of Him who sent Me."* (John 5:22, 30)

- *And He who searches the hearts knows what the mind of the Spirit is, because He intercedes for the saints according to the will of God.* (Romans 8:27)

2. According to Romans 8:1, what is **not** the reason for God's inspection of the believer?
   - *Therefore, there is now no condemnation for those who are in Christ Jesus.*

   Likewise, when a military inspection occurs, the end goal is for improvement, not for condemnation. Pride and competition are used to motivate the person and the unit to be the best they can possibly be. Because pride is involved, it may not always feel like one is not being condemned, but any person in charge who is not malicious truly wants the unit and each person in the unit to be the best they can be. This desire is pride-induced because it is **their** unit, and their unit's reputation reflects on them. It is also a matter of self-preservation because each person in their unit needs to be combat ready and needs to work well within the unit so that the entire unit is able to function like a well-oiled machine in order to succeed in their mission with as few casualties as possible.

3. What do the following verses tell us about judging (inspecting) others?

   - *Do not judge, and you will not be judged; and do not condemn, and you will not be condemned; pardon, and you will be pardoned. Give, and it will be given to you. They will pour into your lap a good measure – pressed down, shaken together, and running over. For by your standard of measure it will be measured to you in return.* (Luke 6:37-38)

   It may appear that we are never to judge anyone, but when taken in context and compared with other verses in the Bible, that is not a correct interpretation because it is incomplete. Continue researching before you come to a conclusion on this matter. In the following verses, what do you learn about the Christian and judging?

- *But you, why do you judge your brother? Or you again, why do you regard your brother with contempt? For we will all stand before the judgment seat of God.* (Romans 14:10)

- *Does any one of you, when he has a case against his neighbor, dare to go to law before the unrighteous and not before the saints? Or do you not know that the saints will judge the world? If the world is judged by you, are you not competent to constitute the smallest law courts? Do you not know that we will judge angels? How much more matters of this life?* (1 Corinthians 6:1-4)

    - If you continue reading through verse 8 of 1 Corinthians 6, you should realize that this section is more about attitude than judging. The Corinthians are being told by Paul that it is better to allow oneself to be wronged than to attack a brother/sister in Christ, especially with the world watching. However, I have asked you to look at verses 1-4 so that you will realize that not only is it not wrong to judge, but we will judge the world and angels. The issue is our hearts and attitudes when we judge.

4. There was a specific example of inspection which Paul did and called upon the Corinthian church to do. What do you learn about it from the following verses?

- *I wrote you in my letter not to associate with immoral people; I did not at all mean with the immoral people of this world, or with the covetous and swindlers, or with idolaters, for then you would have to go out of the world. But actually, I wrote to you not to associate with any so-called brother if he is an immoral person, or covetous, or an idolater, or a reviler, or a drunkard, or a swindler – not even to eat with such a one. For what have I to do with judging outsiders? Do you not judge those who are within the church? But those who are outside, God judges. Remove the wicked man from among yourselves.* (1 Corinthians 5:9-13)

- *But I determined this for my own sake, that I would not come to you in sorrow again. For if I cause you sorrow, who then makes me glad but the one whom I made sorrowful? This is the very thing I wrote you, so that when I came, I would not have sorrow from those who ought to make me rejoice; having confidence in you all that my joy would be the joy of you all. For out of much affliction and anguish of heart I wrote to you with many tears; not so that you would be made sorrowful, but that you might know the love which I have especially for you. But if any has caused sorrow, he has caused sorrow not to me, but in some degree – in order not to say too much – to all of you. Sufficient for such a one is this punishment which was inflicted by the majority, so that on the contrary you should rather forgive and comfort him, otherwise such a one might be overwhelmed by excessive sorrow. Wherefore I urge you to reaffirm your love for him. For to this end also I wrote, so that I might put you to the test, whether you are obedient in all things. But one whom you forgive anything, I forgive also; for indeed what I have forgiven, if I have forgiven anything, I did it for your sakes in the presence of Christ, so that no advantage would be taken of us by Satan, for we are not ignorant of his schemes.* (2 Corinthians 2:1-11)

5. What insights do you glean from the following verses in the context of Paul's writings to the Corinthian church above?

- *Therefore, let us not judge one another anymore, but rather determine this – not to put a stumbling block in a brother's way.* (Romans 14:13)

- *Do not speak against one another, brethren. He who speaks against a brother or judges his brother, speaks against the law and judges the law; but if you judge the law, you are not a doer of the law but a judge of it. There is only one Lawgiver and Judge, the One who is able to save and to destroy; but who are you who judge your neighbor?* (James 4:11-12)

6. What other kind of inspections (judging) should we do?

- *Therefore, whoever eats the bread or drinks the cup of the Lord in an unworthy manner, shall be guilty of the body and the blood of the Lord. But a man must examine himself, and in so doing he is to eat of the bread and drink of the cup. For he who eats and drinks, eat and drinks judgment to himself if he does not judge the body rightly. For this reason, many among you are weak and sick, and a number sleep. But if we judged ourselves rightly, we would not be judged. But when we are judged, we are disciplined by the Lord so that we will not be condemned along with the world.* (1 Corinthians 11:27-32)

There are many verses about not judging, yet there are some that tell us to judge. Is Scripture contradicting itself? No. Scripture never contradicts itself. We just need to search more to understand the context and how these various verses give us a fuller picture. From my understanding, our judging is never for condemnation, but for improvement purposes. Always, our judgements should be made in love and with the understanding that we, too, can easily be led astray, fail, and fall. We judge in order to correct ourselves. If a person is not a believer, we judge them and ourselves in order to determine if they have a positive or negative influence on us and if we are able to influence them in a positive way. Godly judging on our part does not entail us looking down on the one we judge, for we realize that "There, but for the grace of God, go I." We judge others only for the purpose of helping them know Christ or grow in Christ, or for keeping ourselves from temptations that would harm our relationship with Christ.

Too many believers (in so-called "high" and "low" positions) judge with wrong motives, without humility, or without a correct understanding of the Scriptures. Additionally, many non-believers or Christians who are comfortably living with a sin love to scare others away from judging by quoting, "Judge not, lest you be judged" without considering the context of this statement and its relationship with other verses. Do not be scared away from judging; just judge rightly, with love and with a pure heart, and with the goal of helping strengthen your brothers and sisters in the faith.

Chapter 7: History

There are many who do not understand the importance of learning history. I fear the main reason for this is that we have had far too many history teachers who made history boring and who didn't include ways to understand why it was important to understand history. There have been times I have wondered what it would be like for me to teach history. I believe the most important aspects of history are: what led up to the event, what was the event, and what was the result of the event. Names and dates, in my opinion, are only important for understanding the setting for that event in history as compared with other historical events. What would it be like if we were taught with a focus on what I believe are the more important aspects instead of the settings?

In basic training, recruits learn about specific historical events (usually battles) in which their branch of service was involved. In the Marine Corps, we learned how the Continental Marines were created on November 10, 1775, and how closely tied to the Navy we were at the beginning – protecting ships and ports. Particular battles are highlighted, especially those that explain how they brought about our nicknames and others' reactions to us. For example, because of how often swords cut into the necks of soldiers, a leather band was used around the neck in our uniform. This gave us the nickname of Leatherneck. Another example is how we earned the name Devil Dog for the way our 4th brigade fought in the battle of Belleau Wood during WWI.

We also learn about specific people in our history – how they overcame obstacles, how they led and fought, and how they displayed heroism. One Marine example is Lieutenant General Lewis "Chesty" Puller. His leadership skills were exceptional and he led by example, never holding himself to a different standard than the men under him. He was in the horrendous Battle of Chosin Reservoir and is quoted as having said, "We've been looking for the enemy for some time now. We finally found him. We are surrounded. That simplifies things."[1] Another Marine example is Sergeant Major Daniel J. Daly. He was a two-time recipient of the Medal of Honor, which is the highest military award and is given to a military person who has distinguished himself/herself by acts of valor. During a fight when his unit was outnumbered, Sgt. Maj. Daly encouraged his Marines by yelling, "For Christ's sake men, come on! Do you want to live forever?"[2]

The reasons for the history learned in boot camp seems to me to be mostly to bring about comradery and pride, and to encourage desired attributes like heroism and a "can do" attitude. We develop comradery because once recruits graduate, they all now share the same history. This gives each Marine a sense of self. We ARE our history. In fact, even many years after leaving active service in the Marine Corps, as soon as someone I meet and I learn we are both Marines, instantly we feel a special bond. The pride gained through learned history is because of all the heroic things those before us have done as well as how groups of our military branch succeeded in their missions. Training changes as the years go by; however, the core of

---

[1] Brown, Sean Mclain. "5 Reasons Why Chesty Puller Is a Marine Corps Legend." Military.com, 2024, <www.military.com/marine-corps/5-reasons-why-chesty-puller-marine-corps-legend.html> Feb. 2024

[2] "Daniel Joseph Daly." Mt. Soledad National Veterans Memorial, <https://soledadmemorial.org/plaques/sergeant-major-daniel-joseph-daly/> Nov. 2024

what we share remains the same. In fact, a young female active Marine displayed discernment and honor when she told me (an old female former Marine), a very important fact, saying that it was because of the women Marines that went before her that she was able to do what she was currently doing in the Corps, and she valued us and our sacrifices greatly. This makes me think of 1 Corinthians 3:10, "According to the grace of God which was given to me, like a wise master builder I laid a foundation, and another is building on it. But each man must be careful how he builds on it."

For the Christian, the Bible covers all we need to know for life on earth as a Believer, for "*All Scripture is God-breathed and is useful for teaching, rebuking, correcting and training in righteousness*" (2 Timothy 3:16). The Bible is more than our history book, but for now, let's just look at how some examples of the history it shows are beneficial to us.

Looking at the history shown within the pages of the Bible, we can also focus on events and people, but we can also see God's hand more clearly than in the US military history (though it is very obvious at the beginning of our country). I won't get into too many specifics; instead, I suggest that you read and study your Bible to learn more.

Some events which I would like to highlight are mostly concerning Israel because God has chosen to use Israel to show the world who He is and how He wants to relate to us. Nevertheless, these examples should bring you pride in the Lord God and encouragement knowing that you are His. They should also reveal the attributes in people that God uses, especially the attributes He wants us to have.

The Exodus

1. *But Moses said to the people, "Do not fear! Stand by and see the salvation of the Lord which He will accomplish for you today; for the Egyptians whom you have seen today, you will never see them again forever. The Lord will fight for you while you keep silent."*

   Then the Lord said to Moses, "Why are you crying out to Me? Tell the sons of Israel to go forward. As for you, lift up your staff and stretch out your hand over the sea and divide it, and the sons of Israel shall go through the midst of the sea on dry land. As for Me, behold, I will harden the hearts of the Egyptians so that they will go in after them; and I will be honored through Pharaoh and all his army, through his chariots and his horsemen. Then the Egyptians will know that I am the Lord, when I am honored through Pharaoh, through his chariots and his horsemen."

   At the morning watch, the LORD looked down on the army of the Egyptians through the pillar of fire and cloud and brought the army of the Egyptians into confusion. He caused their chariot wheels to swerve, and He made them drive with difficulty; so the Egyptians said, "Let us flee from Israel, for the LORD is fighting for them against the Egyptians." (Exodus 14:13-20, 24-25)

   - What 4 things in the first paragraph (vs 13-14), are the people told to do?

-
-
-
-

- After this, will the Egyptians fear Israel because of the power of the Israelites? Explain your answer.

2. *Thus, the LORD saved Israel that day from the hand of the Egyptians, and Israel saw the Egyptians dead on the seashore. When Israel saw the great power which the LORD had used against the Egyptians, the people feared the LORD, and they believed in the LORD and in His servant Moses.* (Exodus 14:30-31)

    - How did the Israelites react to these events?

    - Consider how you would react if you had been an Israelite there at that time. Would what you just saw and experienced create a common bond with others who had experienced the same thing? Would you have more pride in your God, whose strength you had just witnessed? Would you have more trust in your leader (Moses) through whom God worked these miracles?

    - Now, consider who you are in relation to God. (*"For God so loved the world, that He gave His only begotten Son, that whoever believes in Him shall not perish, but have eternal life."* John 3:16) Can you have the same feelings of pride in God and unity with others who have the same relationship with God? Why/Why not?

Later in Exodus, we learn about how God created a cohesiveness with the Israelites. He made a covenant with them, gave them laws to govern themselves by, and gave them a unifying method of worshipping Him. They all, each one of them, fell under these instructions and this contract. This unified them and, though many have tried to break up that unity over the centuries, they are still unified by these laws, the same religious culture, and this covenant.

The Early Church

3. *So, when they had come together, they were asking Him, saying, "Lord is it at this time You are restoring the kingdom to Israel?" He said to them, "It is not for you to know the times or epochs which the Father has fixed by His own authority; but you will receive power when the Holy Spirit has come upon you; and you shall be My witnesses both in Jerusalem, and in all Judea and Samaria, and even to the remotest part of the earth."* (Acts 1:6-8)

    - The disciples' focus was on Israel at this time. That was with whom they had a shared history, culture, and religion. To where did Jesus want them to shift their focus?

4. *When the day of Pentecost had come, they were all together in one place. And suddenly there came from heaven a noise like a violent rushing wind, and it filled the whole house where they were sitting. And there appeared to them tongues as of fire distributing themselves, and they rested on each one of them. And they were all filled with the Holy Spirit and began to speak with other tongues, as the Spirit was giving them utterance.*

    Now there were Jews living in Jerusalem, devout men from every nation under heaven. And when this sound occurred the crowd came together, and were bewildered because each one of them was hearing them speak in his own language. (Acts 2:1-6)

    - What nationality & religion was the crowd?

      They were still of the same religious background - Judaism, but they did not likely have all the same customs because they came from different nations. Therefore, this was God's first step in expanding his Church.

5. *Saul was in hearty agreement with putting him to death. And on that day a great persecution began against the church in Jerusalem, and they were all scattered throughout the regions of Judea and Samaria, except the apostles.... Therefore, those who had been scattered went about preaching the word. Philip went down to the city of Samaria and began proclaiming Christ to them.* (Acts 8:1, 4-5)

    - What caused the church to begin to expand beyond the Jewish believers?

- What new thing did the believers now have in common?

- Have you and another person ever gone through some difficulty together? How did it affect your relationship?

- Have you ever met someone who has experienced the same kind of experience you have? How do you feel as soon as you learn that?

The difficulties experienced through basic training create a comradery. A shared history and shared experiences (especially difficult ones) will bring about a special connection, a type of bonding, almost immediately. That's human nature. That's how God made us. Therefore, let us follow the instructions of Hebrews 10:24-25 and build those connections.

> *And let us consider how to stimulate one another to love and good deeds, not forsaking our own assembling together, as is the habit of some, but encouraging one another; and all the more as you see the day drawing near.* (Hebrews 10:24-25)

The Post-Apostolic Church

After the original twelve apostles had died, the Church went through several seasons of change. You can get a good idea of the various seasons of change by reading chapters 2 and 3 of the Book of Revelation. These two chapters specifically mention churches which were functioning when the apostle John was alive, yet they had issues the Church has dealt with for its entire life. You can also learn more through the two documentaries *The 7 Churches of Revelation: Times of Deception* and *The 7 Churches of Revelation: Times of Fire*. There is so much history of the Church, both in the Bible and extra-biblical, that I will not get into it here, for that would take another book or more. Instead, I challenge you to read about the Church and various historical figures which had great effect on her. Those true stories can provide a firmer understanding of what the Church has gone through, how Christians' actions affected her and affected non-believers, and how God has been working with and through the Church over the years. You will also learn right and wrong ways to behave, especially since you'll be able to see the outcome of such actions.

.

Chapter 8: The Eleven General Orders

It was one night during my last year in the US Marine Corps that the Lord showed me correlations between the 11 General Orders and our Christian walk. These orders are memorized in boot camp. One way they are personalized and rooted deep within us is that we always begin with, *"My ___* [first, second, etc.] *General Order is...."* These General Orders are a guide for every time a service member is on guard duty, though some could include other times. As Christians, guard duty is a never-ending need; therefore, the spiritual side of these General Orders should be taken very seriously and followed closely.

1. My first General Order is *"To take charge of this post and all government property in view."*[1]

   ➢ What does it mean to *"take charge of"* them?

   - In the military, taking charge of these things means that I am responsible for them. I will protect them from harm and from thievery. In order to do this, I must position myself to be able to see them and I must be ready to physically defend them.
   - In the spiritual, taking charge of these things is the same as taking responsibility to care for them. We take care of them physically through our actions, but we also take care of them spiritually through prayer. We must position ourselves to have knowledge and understanding through study of Scripture. Additionally, we must be ready to defend them through watchful prayer.

   ➢ *"This post,"* is the individual's body and spirit. Write your insights about taking care of your body and spirit from the following verses.

   - *So, remove grief and anger from your heart and put away pain from your body, because childhood and the prime of life are fleeting.* (Ecclesiastes 11:10)

   - *If your right eye makes you stumble, tear it out and throw it from you; for it is better for you to lose one of the parts of your body, than for your whole body to be thrown into hell. If your right hand makes you stumble, cut it off and throw it from you; for it is better for you to lose one of the parts of your body, than for your whole body to go into hell.* (Matthew 5:29-30)

---

[1] USMC Marine Corps Institute, "The United States Marines – Essential Subjects" (Arlington: USMC, Dept. of the Navy, 1983) page 4-1, and additional General Orders are on pages 4-2 through 4-7. *This source is the same for the remaining General Orders cited in this chapter, so it will not be repeated elsewhere.*

- *The eye is the lamp of the body; so then if your eye is clear, your whole body will be full of light. But if your eye is bad, your whole body will be full of darkness. If then the light that is in you is darkness, how great is the darkness!* (Matthew 6:22-23)

- *Therefore, do not let sin reign in your mortal body so that you obey its lusts, and do not go on presenting the members of your body to sin as instruments of unrighteousness; but present yourselves to God as those alive from the dead, and your members as instruments of righteousness to God.* (Romans 6:12-13)

- *I am speaking in human terms because of the weakness of your flesh. For just as you presented your members as slaves to impurity and to lawlessness, resulting in further lawlessness, so now present your members as slaves to righteousness, resulting in sanctification.* (Romans 6:19)

- *Do you not know that you are a temple of God and that the Spirit of God dwells in you? If any man destroys the temple of God, God will destroy him, for the temple of God is holy, and that is what you are.* (1 Corinthians 3:16-17)

- *Flee immorality. Every other sin that a man commits is outside the body, but the immoral man sins against his own body. Or do you not know that your body is a temple of the Holy Spirit who is in you, whom you have from God, and that you are not your own? For you have been bought with a price: therefore glorify God in your body.* (1 Corinthians 6:18-20)

- *Do not be bound together with unbelievers; for what partnership have righteousness and lawlessness, or what fellowship has light with darkness? Or what harmony has Christ with Belial, or what has a believer in common with an unbeliever? Or what agreement has the temple of God with idols? For we are the temple of the living God; just as God said, "I will dwell in them and walk among them and I will be their God, and they shall be my people."* (2 Corinthians 6:14-16)

- *For no one ever hated his own flesh, but nourishes and cherishes it, just as Christ also does the church, because we are members of His body.* (Ephesians 5:29-30)

- The "*government property in view,*" is the Church body and its individual members. This means the church you attend, but also the larger Church around the world. Just as take charge of *this post* is equivalent to take care of it, so it is for the *government property in view*. Much of what can be discovered from the previous verses for "*this post*" can also be used for the Church and its individual members. What do the following verses say about how we are to take care of the Church body and its members?

  - *If your brother sins, go and show him his fault in private; if he listens to you, you have won your brother. But if he does not listen to you, take one or two more with you, so that by the mouth of two or three witnesses every fact may be confirmed. If he refuses to listen to them, tell it to the church; and if he refuses to listen even to the church, let him be to you as a Gentile and a tax collector.* (Matthew 18:15-17)

  - *So the twelve summoned the congregation of the disciples and said, "It is not desirable for us to neglect the word of God in order to serve tables. Therefore, brethren, select from among you seven men of good reputation, full of the Spirit and of wisdom, whom we may put in charge of this task. But we will devote ourselves to prayer and to the ministry of the word."* (Acts 6:2-4)

  - *So Peter was kept in the prison, but prayer for him was being made fervently by the church to God.* (Acts 12:5)

  - *Be on guard for yourselves and for all the flock, among which the Holy Spirit has made you overseers, to shepherd the church of God which He purchased with His own blood.* (Acts 20:28)

  - *Now I urge you, brethren, by our Lord Jesus Christ and by the love of the Spirit, to strive together with me in your prayers to God for me,* (Romans 15:30)

  - *So also you, since you are zealous of spiritual gifts, seek to abound for the edification of the church.* (1 Corinthians 14:12)

  - *But God has so composed the body, giving more abundant honor to that member which lacked, so that there may be no division in the body, but that the members may have the same care for one another. And if one member suffers, all the members suffer with it; if one member is honored, rejoice with it.* (1 Corinthians 12:24b-26)

- *Apart from such external things, there is the daily pressure on me of concern for all the churches. Who is weak without my being weak? Who is led into sin without my intense concern?* (2 Corinthians 11:28-29)

- *For we rejoice when we ourselves are weak but you are strong; this we also pray for, that you be made complete.* (2 Corinthians 13:9)

- *For this reason I too, having heard of the faith in the Lord Jesus which exists among you and your love for all the saints, do not cease giving thanks for you, while making mention of you in my prayers; that the God of our Lord Jesus Christ, the Father of glory, may give to you a spirit of wisdom and of revelation in the knowledge of Him.* (Ephesians 1:15-17)

- *Therefore I, the prisoner of the Lord, implore you to walk in a manner worthy of the calling with which you have been called, with all humility and gentleness, with patience, showing tolerance for one another in love, being diligent to preserve the unity of the Spirit in the bond of peace.* (Ephesians 4:1-3)

- *And this I pray, that your love may abound still more and more in real knowledge and all discernment, so that you may approve the things that are excellent, in order to be sincere and blameless until the day of Christ;* (Philippians 1:9-10)

- *We proclaim Him, admonishing every man and teaching every man with all wisdom, so that we may present every man complete in Christ.* (Colossians 1:28)

- *Epaphras, who is one of your number, a bondslave of Jesus Christ, sends you his greetings, always laboring earnestly for you in his prayers, that you may stand perfect and fully assured in all the will of God.* (Colossians 4:12)

- *And let us consider how to stimulate one another to love and good deeds, not forsaking our own assembling together, as is the habit of some, but encouraging one another; and all the more as you see the day drawing near.* (Hebrews 10:24-25)

➢ The *"government property in view,"* is also the world and the people in it. Though we cannot take care of them the way we do our brothers and sisters in the Lord, we are still required to *"take charge of"* them. The following are verses concerning this concept. What do you learn from them about how or why we should take care of the world and its people (those who aren't a part of God's army)?

- *The earth is the LORD's, and all it contains, the world, and those who dwell in it.* (Psalm 24:1)

- *But I say to you, love your enemies and pray for those who persecute you, so that you may be sons of your Father who is in heaven; for He causes His sun to rise on the evil and the good, and sends rain on the righteous and the unrighteous. For if you love those who love you, what reward do you have? Do not even the tax collectors do the same? If you greet only your brothers, what more are you doing than others? Do not even the Gentiles do the same? Therefore you are to be perfect, as your heavenly Father is perfect.* (Matthew 5:44-48)

- *And Jesus came up and spoke to them, saying, "All authority has been given to Me in heaven and on earth. Go therefore and make disciples of all the nations, baptizing them in the name of the Father and the Son and the Holy Spirit, teaching them to observe all that I commanded you; and lo, I am with you always, even to the end of the age.* (Matthew 28:18-20)

- *When the scribes of the Pharisees saw that He was eating with the sinners and tax collectors, they said to His disciples, "Why is He eating and drinking with tax collectors and sinners?" And hearing this, Jesus said to them, "It is not those who are healthy who need a physician, but those who are sick; I did not come to call the righteous, but sinners."* (Mark 2:16-17)

- *Brethren, my heart's desire and my prayer to God for them is their salvation.* (Romans 10:1)

- *How then will they call on Him in whom they have not believed? How will they believe in Him whom they have not heard? And how will they hear without a preacher? How will they preach unless they are sent? Just as it is written, "How beautiful are the feet of those who bring good news of good things!"* (Romans 10:14-15)

- *Conduct yourselves with wisdom toward outsiders, making the most of the opportunity. Let your speech always be with grace, as though seasoned with salt, so that you will know how you should respond to each person.* (Colossians 4:5-6)

- *To sum up, all of you be harmonious, sympathetic, brotherly, kindhearted, and humble in spirit; not returning evil for evil or insult for insult, but giving a blessing instead; for you were called for the very purpose that you might inherit a blessing.* (1 Peter 3:9) Though the context of this verse is for wives, husbands, and children in believing families, if we are like it says within our families, shouldn't we be like this all the time?

2. My second General Order is "To walk my post in a military manner, keeping always on the alert and observing everything that takes place within sight or hearing."

   ➢ Walking in a military manner is, for the Christian, behaving in a Christlike manner. How do the following verses describe walking Christlike?

   - *Therefore, be imitators of God, as beloved children; and walk in love, just as Christ also loved you and gave himself up for us, an offering and a sacrifice to God as a fragrant aroma.* (Ephesians 5:1-2)

   - *Only conduct yourselves in a manner worthy of the gospel of Christ, so that whether I come and see you or remain absent, I will hear of you that you are standing firm in one spirit, with one mind striving together for the faith of the gospel;* (Philippians 1:27)

   - *Therefore, as you have received Christ Jesus the Lord, so walk in Him.* (Colossians 2:6)

- *The one who says he abides in Him ought himself to walk in the same manner as He walked.* (1 John 2:6)

➢ We are to keep alert, which means our spiritual eyes should be open in order to decide the best thing to do (pray, warn, or command). What do you glean from the following verses about this command?

- *Let your eyes look directly ahead and let your gaze be fixed straight in front of you. Watch the path of your feet and all your ways will be established. Do not turn to the right nor to the left; turn your foot from evil.* (Proverbs 4:25-27)

- *Keep watching and praying that you may not enter into temptation; the spirit is willing, but the flesh is weak.* (Matthew 26:41)

- *With all prayer and petition pray at all times in the Spirit, and with this in view, be on the alert with all perseverance and petition for all the saints.* (Ephesians 6:18)

- *Devote yourselves to prayer, keeping alert in it with an attitude of thanksgiving.* (Colossians 4:2)

- *Preach the word; be ready in season and out of season; reprove, rebuke, exhort, with great patience and instruction.* (2 Timothy 4:2)

- *Therefore, since we have so great a cloud of witnesses surrounding us, let us also lay aside every encumbrance and the sin which so easily entangles us, and let us run with endurance the race that is set before us, fixing our eyes on Jesus, the author and perfecter of faith.* (Hebrews 12:1-2)

- *Be of sober spirit, be on the alert. Your adversary, the devil, prowls around like a roaring lion, seeking someone to devour.* (1 Peter 5:8)

3. My third General Order is "To report all violations of orders I am instructed to enforce"

- As Christians, we need to be aware of and flee from sin. Reporting such sin involves prayer about and confession of sins (personal, church, country, generational). What do you learn from the following verses about this action?

    - *And I said, "O my God, I am ashamed and embarrassed to lift up my face to you, my God, for our iniquities have risen above our heads and our guilt has grown even to the heavens. Since the days of our fathers to this day we have been exceedingly guilty, for on account of our iniquities we, our kings and our priests have been given into the hand of the kings of the lands, to the sword, to captivity and to plunder and to open shame as it is this day."* (Ezra 9:6-7)

    - *Let your ear now be attentive and Your eyes open to hear the prayer of Your servant which I am praying before You now, day and night, on behalf of the sons of Israel Your servants, confessing the sins of the sons of Israel which we have sinned against You; I and my father's house have sinned. We have acted very corruptly against You and have not kept the commandments, nor the statutes, nor the ordinances which You commanded your servant Moses.* (Nehemiah 1:6-7)

    - *I acknowledged my sin to You, and my iniquity I did not hide; I said, "I will confess my transgressions to the LORD"; and You forgave the guilt of my sin.* (Psalm 32:5)

    - *I prayed to the LORD my God and confessed and said, "Alas, O Lord, the great and awesome God, who keeps His covenant and lovingkindness for those who love Him and keep His commandments, we have sinned, committed iniquity, acted wickedly and rebelled, even turning aside from your commandments and ordinances. Moreover, we have not listened to Your servants the prophets, who spoke in Your name to our kings, our princes, our fathers and all the people of the land."...Now while I was speaking and praying, and confessing my sin and the sin of my people Israel and presenting my supplication before the Lord my God in behalf of the holy mountain of my God..."* (Daniel 9:4-6, 20)

    - *If we confess our sins, He is faithful and righteous to forgive us our sins and to cleanse us from all unrighteousness.* (1 John 1:9)

4. My fourth General Order is "To repeat all calls from posts more distant from the guardhouse than my own."

- Repeating a call is passing on the received information so that it gets to those who need to hear it and in order that correct action can be taken. This is equivalent to asking for prayer and praying with others, for the guardhouse is where the Sergeant of the Guard (Jesus or God) would be. What insights do you get from the following verses about this?

  - *Again I say to you, that if two of you agree on earth about anything that they may ask, it shall be done for them by My Father who is in heaven. For where two or three have gathered together in My name, I am there in their midst.* (Matthew 18:19-20)

  - *These all with one mind were continually devoting themselves to prayer, along with the women, and Mary the mother of Jesus, and with His brothers.* (Acts 1:14)

  - *So Peter was kept in the prison, but prayer for him was being made fervently by the church to God.* (Acts 12:5)

  - *With all prayer and petition pray at all times in the Spirit, and with this in view, be on the alert with all perseverance and petition for all the saints, and pray on my behalf, that utterance may be given to me in the opening of my mouth, to make known with boldness the mystery of the gospel.* (Ephesians 6:18-19)

  - *First of all, then, I urge that entreaties and prayers, petitions and thanksgivings, be made on behalf of all men, for kings and all who are in authority, so that we may lead a tranquil and quiet life in all godliness and dignity.* (1 Timothy 2:1-2)

5. My fifth General Order is "To quit my post only when properly relieved."

- A person on guard duty is to stay until another comes to replace him or until he is told by the proper authority that he can leave. Likewise, Christians should never let up unless God says to. What do the following verses show about this?

  - *And they said to Moses, "The people are bringing much more than enough for the construction work which the Lord commanded us to perform." So Moses issued a command, and a proclamation was circulated throughout the camp, saying, "Let no man or woman any longer perform work for the contributions of the sanctuary." Thus the people were restrained from bringing any more. For the material they had was sufficient and more than enough for all the work, to perform it.* (Exodus 36:5-7)

- *Now He was telling them a parable to show that at all times they ought to pray and not to lose heart.* (Luke 18:1)

- *They passed through the Phrygian and Galatian region, having been forbidden by the Holy Spirit to speak the word in Asia; and after they came to Mysia, they were trying to go into Bithynia, and the Spirit of Jesus did not permit them.* (Acts 16:6-7)

- *Therefore, my beloved brethren, be steadfast, immovable, always abounding in the work of the Lord, knowing that your toil is not in vain in the Lord.* (1 Corinthians 15:58)

- *Let us not lose heart in doing good, for in due time we will reap if we do not grow weary.* (Galatians 6:9)

- *I press on toward the goal for the prize of the upward call of God in Christ. Let us therefore, as many as are perfect, have this attitude; and if in anything you have a different attitude, God will reveal that also to you; however, let us keep living by that same standard to which we have attained.* (Philippians 3:14-16)

- *But as for you, brethren, do not grow weary of doing good.* (2 Thessalonians 3:13)

- *Therefore, since we have so great a cloud of witnesses surrounding us, let us also lay aside every encumbrance and the sin which so easily entangles us, and let us run with endurance the race that is set before us, fixing our eyes on Jesus, the author and perfecter of faith.* (Hebrews 12:1-2)

- *Blessed is a man who perseveres under trial; for once he has been approved, he will receive the crown of life which the Lord has promised to those who love Him.* (James 1:12)

6. My sixth General Order is "To receive, obey, and pass on to the sentry who relieves me all orders from the Commanding Officer, Officer of the Day, and Officers and Noncommissioned Officers of the Guard only."

   - Christians receive their orders and instructions from the Lord through His Word and through fellowship with God and other Believers. They are then called to live out (obey) and teach others (pass on) what they have learned. The following verses explain all this. Write any specific information or explanation you learn on why or how we are to fulfill this command.

     - *These words, which I am commanding you today, shall be on your heart. You shall teach them diligently to your sons and shall talk of them when you sit in your house and when you walk by the way and when you lie down and when you rise up.* (Deuteronomy 6:6-7)

     - *Train up a child in the way he should go, even when he is old he will not depart from it.* (Proverbs 22:6)

     - *Go therefore and make disciples of all nations, baptizing them in the name of the Father and the Son and the Holy Spirit, teaching them to observe all that I commanded you; and lo, I am with you always, even to the end of the age.* (Matthew 28:19-20)

     - *Let the word of Christ richly dwell within you, with all wisdom teaching and admonishing one another with psalms and hymns and spiritual songs, singing with thankfulness in your hearts to God. Whatever you do in word or deed, do all in the name of the Lord Jesus, giving thanks through Him to God the Father.* (Colossians 3:16-17)

     - *The things which you have heard from me in the presence of many witnesses, entrust these to faithful men who will be able to teach others also.* (2 Timothy 2:2)

     - *But sanctify Christ as Lord in your hearts, always being ready to make a defense to everyone who asks you to give an account for the hope that is in you, yet with gentleness and reverence.* (1 Peter 3:15)

7. My seventh General Order is "To talk to no one except in the line of duty."

    - If it doesn't pertain to one's duty, talking with someone else will take one's mind away from the job at hand. This would be like "fraternizing" with the world for the Believer. What do the following verses say about this order?

        - *And do not be conformed to this world, but be transformed by the renewing of your mind, so that you may prove what the will of God is, that which is good and acceptable and perfect.* (Romans 12:2)

        - *Do not be bound together with unbelievers; for what partnership have righteousness and lawlessness, or what fellowship has light with darkness? Or what harmony has Christ with Belial, or what has a believer in common with an unbeliever?* (2 Corinthians 6:14-15)

        - *For am I now seeking the favor of men, or of God? Or am I striving to please men? If I were still trying to please men, I would not be a bond-servant of Christ.* (Galatians 1:10)

        - *See to it that no one takes you captive through philosophy and empty deception, according to the tradition of men, according to the elementary principles of the world, rather than according to Christ.* (Colossians 2:8)

        - *Suffer hardship with me, as a good soldier of Christ Jesus. No soldier in active service entangles himself in the affairs of everyday life, so that he may please the one who enlisted him as a soldier.* (2 Timothy 2:3-4)

        - *You adulteresses, do you not know that friendship with the world is hostility toward God? Therefore, whoever wishes to be a friend of the world makes himself an enemy of God.* (James 4:4)

        - *As obedient children, do not be conformed to the former lusts which were yours in your ignorance, but like the Holy One who called you, be holy yourselves also in all your behavior.* (1 Peter 1:14-16)

- *Do not love the world nor the things in the world. If anyone loves the world, the love of the Father is not in him. For all that is in the world, the lust of the flesh and the lust of the eyes and the boastful pride of life, is not from the Father, but is from the world. The world is passing away, and also its lusts; but the one who does the will of God lives forever.* (1 John 2:15-17)

8. My eighth General Order is "To give the alarm in case of fire or disorder."

➢ If there is a problem, the guard must react quickly so that correct action can be taken and the problem dealt with before it gets severe. For the Christian, this means we need to tell others of urgent prayer needs and give warnings to those in danger (of slipping, of false doctrine, from a wolf in sheep's clothing, etc.), yet this is not to be done in a judgmental fashion. It's a warning, not a shooting. What do the following verses tell us about giving an alarm?

- *But if the watchman sees the sword coming and does not blow the trumpet and the people are not warned, and a sword comes and takes a person from them, he is taken away in his iniquity; but his blood I will require from the watchman's hand. Now as for you, son of man, I have appointed you a watchman for the house of Israel; so you will hear a message from My mouth and give them a warning from Me.* (Ezekiel 33:6-7)

- *Therefore be on the alert, remembering that night and day for a period of three years I did not cease to admonish each one with tears.* (Acts 20:31)

- *Brethren, even if anyone is caught in any trespass, you who are spiritual, restore such a one in a spirit of gentleness; each one looking to yourself, so that you too will not be tempted.* (Galatians 6:1)

- *Devote yourselves to prayer, keeping alert in it with an attitude of thanksgiving; praying at the same time for us as well, that God will open up to us a door for the word, so that we may speak forth the mystery of Christ, for which I have also been imprisoned.* (Colossians 4:2-3)

- *But you, brethren, are not in darkness, that the day would overtake you like a thief; for you are all sons of light and sons of day. We are not of night nor of darkness; so then let us not sleep as others do, but let us be alert and sober.* (1 Thessalonians 5:4-6)

- *Preach the word; be ready in season and out of season; reprove, rebuke, exhort, with great patience and instruction.* (2 Timothy 4:2)

- *Be of sober spirit, be on the alert. Your adversary, the devil, prowls around like a roaring lion, seeking someone to devour.* (1 Peter 5:8)

9. My ninth General Order is "To call the Corporal of the Guard in any case not covered by instructions."

   ➢ When a guard is posted, he is given instructions; however, sometimes things happen which the instructions don't address. This is when you need to get help for areas you aren't able to handle (due to lack of knowledge or strength). What do the following verses tell you about getting such help?

   - *Two are better than one because they have a good return for their labor. For if either of them falls, the one will lift up his companion. But woe to the one who falls when there is not another to lift him up. Furthermore, if two lie down together they keep warm, but how can one be warm alone? And if one can overpower him who is alone, two can resist him. A cord of three strands is not quickly torn apart.* (Ecclesiastes 4:9-12)

   - *Where there is no guidance the people fall, but in abundance of counselors there is victory.* (Proverbs 11:14)

   - *Without consultation, plans are frustrated, but with many counselors they succeed.* (Proverbs 15:22)

   - *Iron sharpens iron, so one man sharpens another.* (Proverbs 27:17)

- *So the twelve summoned the congregation of the disciples and said, "It is not desirable for us to neglect the word of God in order to serve tables. Therefore, brethren, select from among you several men of good reputation, full of the Spirit and of wisdom, whom we may put in charge of this task. But we will devote ourselves to prayer and to the ministry of the word."* (Acts 6:2-4)

- *Bear one another's burdens, and thereby fulfill the law of Christ.* (Galatians 6:2)

- *Therefore encourage one another and build up one another, just as you also are doing.* (1 Thessalonians 5:11)

10. My tenth General Order is "To salute all Officers and all Colors and Standards not cased."

➢ Colors (flags or banners) and Standards (similar to colors, but usually smaller) symbolize unit identity, tradition, and values. For the US military person, all of this would relate to our country and our Constitution, to which we swore allegiance when we entered the military. These also relate to our particular military branch and the various levels of units we are a part of. Allow me to clarify with an example. The United States Flag ("Old Glory") is referred to as our Colors, and represents the USA, its Constitution, and its basic values and traditions. A Standard could be the flag for a particular unit and would stand for that unit's purpose, successes, traditions, and values. Saluting these symbols is equivalent to giving what they represent honor, reverence, praise, and/or thanksgiving. As Believers, Colors and Standards would represent God (think of the Christian flag), yet we are also to give respect and honor to godly leaders (officers). How do the following verses describe doing this?

- *Ascribe to the LORD the glory due His name; Bring an offering and come before him; Worship the LORD in holy array.* (1 Chronicles 16:29)

- *He who offers a sacrifice of thanksgiving honors Me; And to him who orders his way aright I shall show the salvation of God.* (Psalm 50:23)

- *Come, let us worship and bow down, Let us kneel before the LORD our Maker.* (Psalm 95:6)

- *Enter His gates with thanksgiving and His courts with praise. Give thanks to Him, bless His name.* (Psalm 100:4)

- *But we request of you, brethren, that you appreciate those who diligently labor among you, and have charge over you in the Lord and give you instruction, and that you esteem them very highly in love because of their work. Live in peace with one another.* (1 Thessalonians 5:12-13)

- *Through Him then, let us continually offer up a sacrifice of praise to God, that is, the fruit of lips that give thanks to His name.* (Hebrews 13:15)

- *Obey your leaders and submit to them, for they keep watch over your souls as those who will give an account. Let them do this with joy and not with grief, for this would be unprofitable for you.* (Hebrews 13:17)

- *Rejoice in the Lord always; again I will say, rejoice!* (Philippians 4:4)

- *Now to the King eternal, immortal, invisible, the only God, be honor and glory forever and ever. Amen.* (1 Timothy 1:17)

- *The elders who rule well are to be considered worthy of double honor, especially those who work hard at preaching and teaching.* (1 Timothy 5:17)

- *Worthy are You, our Lord and our God, to receive glory and honor and power; for You created all things, and because of Your will they existed, and were created.* (Revelation 4:11)

11. My eleventh General Order is "To be especially watchful at night and during the time for challenging, to challenge all persons on or near my post, and to allow no one to pass without proper authority."

Remember that "my post" is my body, which includes my mind. This verse for the Christian means guarding my mind, being discerning, not allowing the enemy a foothold in anyone's life (if at all possible and as far as you are able to). Test the spirits. When others are "asleep" — whether spiritually or mentally — it becomes especially challenging to see clearly due to distractions or misinformation. In times like these, or when the days are full of dark deeds, it is crucial to remain especially vigilant.

➢ How do the following verses explain this necessary action?

- *For the weapons of our warfare are not of the flesh, but divinely powerful for the destruction of fortresses. We are destroying speculations and every lofty thing raised up against the knowledge of God, and we are taking every thought captive to the obedience of Christ.* (2 Corinthians 10:4-5)

- *Finally, be strong in the Lord and in the strength of His might. Put on the full armor of God, so that you will be able to stand firm against the schemes of the devil. For our struggle is not against flesh and blood, but against the rulers, against the powers, against the world forces of this darkness, against the spiritual forces of wickedness in the heavenly places.* (Ephesians 6:10-12)

- *Be anxious for nothing, but in everything by prayer and supplication with thanksgiving let your requests be made known to God. And the peace of God, which surpasses all comprehension, will guard your hearts and your minds in Christ Jesus. Finally brethren, whatever is true, whatever is honorable, whatever is right, whatever is pure, whatever is lovely, whatever is of good repute, if there is any excellence and if anything worthy of praise, dwell on these things.* (Philippians 4:6-8)

- *Submit therefore to God. Resist the devil and he will flee from you.* (James 4:7)

- *Be of sober spirit, be on the alert. Your adversary, the devil, prowls around like a roaring lion, seeking someone to devour. But resist him, firm in your faith, knowing that the same experiences of suffering are being accomplished by your brethren who are in the world.* (1 Peter 5:8-9)

- *Beloved, do not believe every spirit, but test the spirits to see whether they are from God, because many false prophets have gone out into the world.* (1 John 4:1-6)

Review of Eleven General Orders as It Pertains to Christians

- **First General Order** p.87 🎖 *To take charge of this post and all government property in view*
  - I am to care for myself (my body, mind, and spirit) by keeping as pure, holy, and healthy as I can.
  - I am to actively seek God's best for the Church and the world through prayer and actions.
- **Second General Order** p.92 🎖 *To walk my post in a military manner, keeping always on the alert and observing everything that takes place within sight or hearing.*
  - I am to be Christlike in conduct and mindset, staying focused on the tasks He has given me.
  - I should keep myself aware of what is happening around me and among my fellow Christians so that I can know how and what to pray, and so that I will know when and how to take action.
- **Third General Order** p.93 🎖 *To* report all violations of orders I am instructed to enforce.
  - Confess personal, group, Church, country and generational sins in prayer (of repentance).
- **Fourth General Order** p.94 🎖 *To repeat calls from posts more distant from the guardhouse than my own.*
  - Pray for others (alone & with others), and ask others to pray for me.
- **Fifth General Order** p.95 🎖 *To quit my post only when properly relieved.*
  - I must never stop doing what God has told me to until He, Himself, tells me to stop.
- **Sixth General Order** p.97 🎖 *To receive, obey, and pass on to the sentry who relieves me all orders from the Commanding Officer, Officer of the Day, and Officers and Noncommissioned Officers of the Guard only.*
  - I receive God's Word and His lessons by accepting them.
  - I obey God's Word and His lessons by altering my lifestyle so that I live them out.
  - I pass on to others God's Word and His lessons by teaching them through my words and actions.
- **Seventh General Order** p.98 🎖 *To talk to no one except in the line of duty.*
  - I must maintain my focus living well for Christ and not allow anything or anyone to distract me from it.
- **Eighth General Order** p.99 🎖 *To give the alarm in case of fire or disorder.*
  - I must let others know of urgent prayer needs.
  - I must do what I can to help other believers not fall into sin or be deceived.
- **Ninth General Order** p.100 🎖 *To call the Corporal of the Guard in any case not covered by instructions.*
  - I don't know everything and can't do everything by myself, so I need to ask for help when needed.
- **Tenth General Order** p.101 🎖 *To salute all Officers and Colors and Standards not cased.*
  - I am to give God honor, glory, and praise.
  - I am to give respect and honor to godly leaders.
- **Eleventh General Order** p.102 🎖 *To be especially watchful at night and during the time of challenging, to challenge all persons on or near my post, and to allow no one to pass without proper authority.*
  - I must guard my mind and not allow anything to take up residence there that is not approved by God.
  - The darker and crazier life around me gets, the more aware and alert I must be for the challenges and distractions that would seek to pull me from full commitment to God.

Chapter 9: After Boot Camp

After graduating from Basic Training, the military person has a different perspective than prior to entering boot camp. He is now to continue in what has been learned, to increase knowledge about how best to live this new lifestyle, to learn more specific information in order to perform the particular jobs assigned, and to act upon all the information gained.

As I stated in the first chapter, Christians must go through Basic Training at the same time as we deal with real spiritual warfare. There is no "safe" training area that prepares us for the real world. Nevertheless, like those in the military, we must continue with what basics we learn, add to our understanding, and comport ourselves accordingly.

1. What do the following verses tell you about how you are to behave in relation to what you have been given, have gained, or have learned?

    - *Take pains with these things; be absorbed in them, so that your progress will be evident to all. Pay close attention to yourself and to your teaching; persevere in these things, for as you do this you will ensure salvation both for yourself and for those who hear you.* (1 Timothy 4:15-16)

    - *You, however, continue in the things you have learned and become convinced of, knowing from whom you have learned them* (2 Timothy 3:14)

    - *Let us not lose heart in doing good, for in due time we will reap if we do not grow weary.* (Galatians 6:9)

    - *Therefore, take up the full armor of God, so that you will be able to resist in the evil day, and having done everything, to stand firm.* (Ephesians 6:13)

    - *So then, my beloved, just as you have always obeyed, not as in my presence only, but now much more in my absence, work out your salvation with fear and trembling.* (Philippians 2:12)

    - *I press on toward the goal for the prize of the upward call of God in Christ Jesus. Let us therefore, as many as are perfect, have this attitude; and if in anything you have a different attitude, God will reveal that also to you; however, let us keep living by that same standard to which we have attained. Brethren, join in following my example, and observe those who walk according to the pattern you have in us. For many walk, of whom I often told you, and now tell you even weeping, that they are enemies of the cross of Christ, whose end is destruction, whose god is their appetite, and whose glory is in their shame, who set their minds on earthly things. For our citizenship is in heaven, from which also we eagerly wait for a Savior, the Lord Jesus Christ; who will transform the body of our humble state into conformity with the body of His glory, by the exertion of the power that He has even to subject all things to Himself. Therefore, my beloved brethren whom I long to see, my joy and crown, in this way stand firm in the Lord, my beloved.* (Philippians 3:14-4:1)

- *And this I pray, that your love may abound still more and more in real knowledge and all discernment,* (Philippians 1:9)

- *Therefore, if you have been raised up with Christ, keep seeking the things above, where Christ is, seated at the right hand of God. Set your mind on the things above, not on the things that are on the earth. For you have died and your life is hidden with Christ in God. When Christ, who is our life, is revealed, then you also will be revealed with Him in glory.*

  Therefore, consider the members of your earthly body as dead to immorality, impurity, passion, evil desire, and greed, which amounts to idolatry. For it is because of these things that the wrath of God will come upon the sons of disobedience, and in them you also once walked, when you were living in them. But now you also, put them all aside: anger, wrath, malice, slander, and abusive speech from your mouth.

  Do not lie to one another, since you laid aside the old self with its evil practices, and have put on the new self who is being renewed to a true knowledge according to the image of the One who created him – a renewal in which there is no distinction between Greek and Jew, circumcised and uncircumcised, barbarian, Scythian, slave and freeman, but Christ is all, and in all.

  So, as those who have been chosen of God, holy and beloved, put on a heart of compassion, kindness, humility, gentleness and patience; bearing with one another, and forgiving each other, whoever has a complaint against anyone; just as the Lord forgave you, so also should you. (Colossians 3:1-13)

- *But grow in the grace and knowledge of our Lord and Savior Jesus Christ. To Him be the glory, both now and to the day of eternity. Amen.* (2 Peter 3:18)

Chapter 10: Summary

Your life as a true, born-again Christian begins the moment you confess your sins to God and repent of them; accept that Jesus is God who came in the flesh, died for your sins, and rose from the dead; and choose for Him to be Lord over your life. (See Romans 3:23, 5:8, 6:23, 10:9-10, and 10:13). This mirrors the Oath of Enlistment in that you are swearing allegiance—not to a country—but to God. Just as military allegiances are lifelong, your commitment to God is eternal. From this point forward, your entire life will be shaped by this new allegiance.

Once you have made this commitment, you acknowledge God as the ultimate authority in your life. In a military Chain of Command, authority flows from the top down. For the Christian, God is at the top of that chain, and no one and nothing else holds higher authority. Although the concept of the triune God is beyond our full understanding, we know that the Father holds ultimate authority, even over the Son, as seen in 1 Corinthians 15:28 and John 14:28. This is why we can refer to God the Father as the Commander in Chief, a title also used by the military for the President of the United States. Jesus, being relatable and having walked among us, can be seen as our Commanding Officer or Captain. Ultimately, both the Father and the Son hold authority over us. The beauty of this relationship, though, is that God is not a distant figure of authority. He invites us into an intimate relationship where we can call Him Father. And as we grow in our relationship with Christ, we look forward to the day when He will no longer just be our Commanding Officer but our Bridegroom (John 3:29, Rev. 19:7-8).

Understanding who you are in Christ is vital right from the start of your walk with Him. The enemy will constantly try to confuse or undermine your identity, knowing that your understanding of who you are will shape how you think and act. This is why, in the military, the first article of the Code of Conduct clearly states who the service member is, their purpose, and the lengths they are willing to go to fulfill that purpose. Translating this to our Christian walk, we might say: "I am a born-again believer, living my life to glorify God and striving to reflect Christ in all I do. I am willing to sacrifice everything, including my life, to make Christ known and to help other believers do the same."

Just as the Code of Conduct goes on to outline the responsibilities of service members, we can adapt its principles to our Christian lives. As believers, we declare: "I will never surrender to the forces of evil, whether external or internal. I will remain aware that my actions reflect on Christ and can either strengthen or weaken my fellow believers. Should I fall, I will strive to free myself from the bondage of sin by relying on the Holy Spirit to guide me back, and I will help others do the same. I will submit to the authority God has placed over me, as long as it aligns with His will as revealed in Scripture. No matter the circumstances, I will continue to trust in God."

Our physical and spiritual health are interconnected, and each one can affect the other. Therefore, we must prioritize both. Additionally, we face a relentless enemy who seeks to "kill, steal, and destroy" (John 10:10). To withstand these attacks, we must learn to administer spiritual "First Aid" to ourselves and others, defend against attacks, and recover from them.

Spiritual readiness involves understanding the enemy's tactics and knowing how to defend ourselves, which is why we must become proficient in spiritual First Aid, NBC Defense, and Fitness.

In the chapter on Uniform Regulations and Clothing, we learned that our "uniform" as Christians is Christ Himself (Rom. 13:14). His character should be evident in our lives, much like insignia on a military uniform is obvious. The "cover" for our mind is Christ's mindset, and over our heart would be the earned ribbons of obedience and growth as we walk with Him.

Much like military boots need regular care and shining, our spiritual lives require ongoing attention. Another routine part of military life is Drill, a continual practice that fosters unity and prompt obedience to commands. In the Christian life, we can think of every experience as a kind of drill. Each situation in life elicits a response, and with practice, we hope to react more quickly and accurately to the commands of our Lord Jesus.

Everyone, whether consciously or not, judges themselves and others, but these Inspections are often flawed or incomplete. In contrast, Christians are called to a higher form of inspection, one that is guided by Scripture and carried out with love and humility. The most important inspection is the one conducted by our loving and forgiving Captain Jesus. Though humans may judge and condemn, only Christ sees with true clarity and holds the authority to condemn or forgive. His evaluations are aimed at growth and improvement, not harm. As Christians, we are called to self-examine, measuring our actions, thoughts, and motivations against Scripture. When we find areas where we fall short, we must take corrective action with the help of God's grace. Additionally, we are sometimes responsible for lovingly inspecting other believers within the church, with the goal of fostering growth, accountability, and restoration. Whether inspecting ourselves or others, these assessments serve to strengthen our faith and our community, helping us to align more closely with Christ. He knows our struggles and lovingly guides us, balancing firm instruction with grace, as we learn to reflect His character more fully.

History provides invaluable lessons for both military personnel and Christians alike. From Biblical history to modern examples, we gain wisdom, courage, and encouragement by learning from those who have gone before us. As Ecclesiastes 1:9 reminds us, "There is nothing new under the sun." By studying how others have handled challenges, we can be better prepared to face our own trials with patience, wisdom, and perseverance.

Tools like history, inspections, and our Code of Conduct equip us to carry out our duties to God and His people. They help us to live out the Eleven General Orders with honor. We take charge of the post God has assigned to us, remain vigilant, communicate with Him, pass on to fellow believers what we have learned, and endure until the Lord relieves us of our duties, not allowing ourselves to be distracted. We encourage and pray for one another, particularly those in leadership, and we strive to glorify God in all we do. In the face of darkness, we remain vigilant, guarding our minds and hearts against the enemy's schemes. These Eleven General Orders guide us until we are relieved of our duty by the Lord calling us home.

Just as boot camp is intentionally designed to challenge and strengthen recruits, life is filled with difficulties that test our resolve and cause us to grow. For the military recruit, there is life After Boot Camp, but for the Christian, boot camp and life happen simultaneously. However, both the Christian and the recruit emerge from their training with a new mindset. Both are tasked with applying the lessons learned in boot camp throughout their lives and passing on what they have learned to others. Living out these values is a lifelong commitment.

Congratulations! You've made it through a challenging and in-depth study—no easy feat. Just as military recruits graduate from their boot camp stronger and better prepared, you have now graduated from this *Spiritual Warfare Boot Camp* if you have made your oath and absorbed these foundational lessons. But remember, just like a soldier's training doesn't stop at graduation, your spiritual journey continues. Now, it's your duty to stand firm in the faith, to apply what you've learned, and to keep building on these basics every day.

Know that what you've gained here equips you to face whatever challenges lie ahead. Even if a situation feels overwhelming, trust that our Commanding Officer—the Almighty Lord—will give you the strength, wisdom, and resources you need when the time comes. Your training is not over; it's ongoing. But you're ready for what's to come because you're not going into battle alone. The Lord Himself ensures you are prepared, and He will never leave you without what you need.

For the several branches of the military, we can call a person who was in the military but is no longer, an "ex" whatever, with one exception – the Marine. In other words, someone can be an ex-Army or ex-Navy person, but never an ex-Marine. Once a Marine, always a Marine. Why is this so? A simplistic way to explain is that the Marine Corps motto is Semper Fidelis, which means "always faithful." Marines are trained with this mindset – we will always be faithful to our country and our Corps. Why do I tell you this? Because the Christian should take on this mindset as well. Christians should always be faithful to their country (heaven) and their Corps (Christians), and mostly to their God (YHVH).

Bibliography

Brown, Sean Mclain. *5 Reasons Why Chesty Puller is a Marine Corps Legend*. n.d. Feb. 2024. <https://www.military.com/marine-corps/5-reasons-why-chesty-puller-marine-corps-legend.html>.

Daniel Joseph Daly. n.d. Nov. 2024. <https://soledadmemorial.org/plaques/sergeant-major-daniel-joseph-daly/>.

The Lockman Foundation. *New American Standard Bible*. La Harba: Shiloh Road Publishers, 1995.

The US Military Oath of Enlistment. 2024. Feb. 2024. <https://www.military.com/join-armed-forces/swearing-in-for-military-service.html>.

USMC Marine Corps Institute. *The United States Marine - Essential Subjects*. Vol. MCIO P1550.14D. Arlington: USMC, Dept of the Navy, 1983.

Author Biography

Cristi Erickson, a U.S. Marine and dedicated Bible teacher, brings a unique perspective to spiritual development through her military experiences. As the daughter of a Navy Mustanger, Cristi grew up traveling across the United States, developing a deep curiosity and a passion for understanding life's complexities. At the age of 13, she gave her life to Christ, and her inquisitive nature, combined with athleticism and a fascination with warfare, set the foundation for her lifelong pursuit of understanding spiritual and physical combat.

Cristi enlisted in the Marine Corps at 21, forgoing her eligibility to become an officer, and served for four years. Recognized early as a leader, she was appointed Guide during boot camp, a role that taught her the value of leading by example and being held to higher standards. Her time in the Corps was marked by discipline, perseverance, and a reliance on God's guidance. Despite planning a military career, Cristi felt called by the Lord to leave the Marines at 25 and pursue a BA in Religion: Cross-Cultural Ministries to prepare for the mission God had for her—wherever He would send her.

Cristi began leading Bible studies in her late teens, an experience that planted the seeds for her growth as a teacher. Over the years, she honed her teaching skills, eventually gaining formal training to teach English to Speakers of Other Languages. Her love for writing is evident in the heartfelt poetry and musings she has shared on social media, which have touched and encouraged many. This workbook, her first published book, reflects her passion for helping others understand spiritual truths through relatable analogies and military-inspired principles.

In *Spiritual Warfare from a Military Perspective: Boot Camp*, Cristi shares lessons learned from her Marine Corps experience, paralleling them with the Christian walk. With a focus on adopting the mind of Christ, working in unity, and standing firm in spiritual battles, her book equips readers with tools to grow in faith and glorify God. Cristi's mission is clear: to inspire others to live effectively as followers of Christ by blending military discipline with spiritual devotion.

Made in the USA
Coppell, TX
05 November 2026

11315635R00066